GOOD
SUPPLY CHAIN
PRACTICES

(G.S.C.P.)

IN THE PHARMACEUTICAL INDUSTRY

" *TOWARDS AGILE MANUFACTURE* "

by Francis GOLDBERGER

PREFACE BY
Sir Richard SYKES

Also by F. Goldberger
Pharmaceutical Manufacturing.

Ebur
Rue Cocherel n° 22
Evreux, 27000
France

Printed in France by the Hérissey Press.

ISBN 2-906016-01-0-2

To
my Family with deep affection and gratitude

The views expressed herein are those of the author alone and neither Glaxo Wellcome nor its subsidiaries, or any other firms mentioned, shall be liable in any manner whatsoever for any matter arising from any statement or errors contained herein.

Please note: at any time in this book where the terms "man" or "men" are applied, they should be considered in their generic sense; thus could refer either to men or to women.

Contents

CHAPTER HEADINGS

ACKNOWLEDGEMENTS

Firstly my grateful thanks to Sir Richard Sykes for having written the Preface to this book.

Then to Glaxo Wellcome p.l.c. and Glaxo Wellcome France; Christopher Viehbacher notably, who have encouraged and helped me in writing this book and allowed me to use some material.

Secondly, I would like to thank Michel De Closets who for 25 years practised the Supply Chain always with the customers in mind. Many examples came out of his rich experience. Thanks also to Ian Brand, Supply Chain Consultant, Glaxo Wellcome Operations at Stockley Park on the section on Information Technology. Leo Lucisano gave me some very helpful notes and criticism. Robert Thompson-Pratt also contributed a lot and edited this book.

Grateful thanks to my Wife who not only edited but also helped me with the grammar and the proof-reading.

Special thanks to Martine Lamiot who tirelessly and patiently wrote and rewrote numerous drafts.

Many thanks to Stephen Blackledge who has a vast knowledge of the Supply Chain, and has greatly contributed to this book.

Imprimerie Hérissey is to be thanked for their patience and professionalism.

Many thanks to my Colleagues for being so supportive, and with whom I have "lived" the Supply Chain; and lastly deep thanks to my customers who for 36 years I strove to satisfy and without whom I could not have accomplished my job.

Preface

by

Sir Richard Sykes

Chairman of Glaxo Wellcome p.l.c.

After a lifetime dedicated to the manufacture and supply of quality pharmaceutical products, there are few people better placed than Francis Goldberger to pass on a valuable body of accumulated wisdom on the topics surveyed in this book.

It crystallises the experience of Francis Goldberger's years of dedication to the challenges and issues of production and supply, and will repay careful study by production and supply-chain managers, and also by students of efficient practice in a modern technology-based industry.

This volume follows on from Francis Goldberger's earlier work, "Pharmaceutical Manufacturing: Quality Management in the Industry" (1991). I have no doubt that it will, like its predecessor, find a wide and enthusiastic readership.

January 1998

Foreword
&
Purpose of book

This book has a number of purposes:
- To situate the supply chain in the context of modern pharmaceutical management thinking and practice, and to give an insight into probable future developments of the industry.
- To provide a concise guide to quality issues in the Supply Chain in the pharmaceutical industry.
- To explain the reasons for the quality issues, often illustrated by examples from the author's 36 years experience in the industry.
- To show the necessity and use of Agility in modern manufacturing.
- To explain the Supply Chain and its overall structure, reason, and organisation, to those in the industry who are connected with the Supply Chain.
- To examine particularly the constraints in the manufacturing environment, both technical, logistical, and regulatory, which could speed up the chain.
- By understanding the above, to improve the Supply Chain, increase performance and overall production efficiency, and reduce cost.
- To revisit and complement the author's book: "Pharmaceutical Manufacturing – Quality Management in the Industry" published in 1991.
- To serve as a text for staff in the manufacturing industry, the suppliers, wholesalers and pharmacists, in order to understand each other's work and challenges.
- To accommodate the Supply Chain in a constantly changing environment, where anticipation and commitment by staff to efficient healthcare is essential for competitive advantage and also the advancement of human well-being.

Francis Goldberger
February 1998

Introduction

Why another Guide to Good Practices?

The Supply Chain and its efficient management is one of the key levers to reduce the cost of manufacture and to improve performance in any industry. This is even more true in the pharmaceutical industry, where raw materials have high costs which are going to increase, and where cycle times and lead times are notoriously long where agility is much required.

There is a general lack of understanding of the Supply Chain in its entirety. Different companies use different terms or concepts for the same thing. Even within one company, terminologies might not be clear.

The chain is strengthened by making sure that all the quality aspects connected to the Supply Chain are properly considered. By understanding the quality issues, one can obtain a much better appreciation of the risks and opportunities involved, thereby permitting a better organisation of the Supply Chain.

In the Pharmaceutical Industry, Supply Chain practice is in its infancy.

As in all new practices, a lot of good as well as bad practices have been tried. In other words it is evolving and will change with experience, with time, and with changing circumstances.

The industry is realising that Supply Chain implementation is vital for developing its business and must be an integral part of its strategy.

The industry itself is evolving very fast and its mentality is changing.

The following are illustrations of this phenomenom:
- it has higher customer expectations.
- patients "pull" from the manufacturer rather than the other way round.

- Research and Development discover what the patient needs and how much the payer is prepared to pay.
- the costs of raw materials are increasing.
- its margins are coming down from the soaring 90 %'s to as low as 35 %. Thus there is an acute awareness, both at the procurement, manufacturing and the marketing levels, of the importance of economies in cost containment.
- improving the Supply Chain performance is the objective, not just cutting costs – "one can't shrink to greatness".
- Mass Customisation is not yet practised within the industry, but agile business techniques are a necessary pathway towards it.
- it outsources more and more manufacturing and supporting functions, this implies more and more subcontractors, suppliers and partners.
- its manufacturing base is consolidated by mergers and take-overs as well as internal rationalisation, thus creating more leverage with suppliers and distributors.
- it is in the process of globalisation, that is to say, that it is sourcing all of the world from all over the world. It has to cater for multiple markets from as much standardised "basic" products as possible. This in turn leads to Local Customisation, which means creating market specific packs from standard products, or at least components.
- it has more and more alliances involving different sourcing routes.
- it has realised that lead times mean not only the process of order to delivery, but order to → cash received.
- it has realised that as far as its suppliers and its distributors are concerned, it is not in a game of adversial competition of company against company. Instead, it is Supply Chains competing against Supply Chains, where reduction of overall cost is in the interest of the Chain.
- it is beginning to share technical and financial risks and responsibility with its suppliers and distributors.

- it has realised that functional barriers between marketing, planning, purchasing, manufacturing, and distribution, must be removed. This will financially optimise the entire supply chain.
- consequently, it has not only to co-ordinate better, but has also to integrate functions, *mind sets,* and *vision.*
- it has realised the necessity of simultaneous and agile information flow which can lead to immediate decisions and hence rapid action.
- shorter life cycles, faster introductions of new products (or line extensions) and specific supply/market channels requirements, not only induce more complexity but require radical factors to achieve success.
- distortions, such as price control in some countries and free flow in others, together with parallel trade and currency fluctuations, contribute to the dynamic complexity.
- it has to be more pro-active and anticipate marketing, financial and technical trends more rapidly.
- flexibility and reactivity have been accepted as current words, but their application is by no means universal or even understood.
- the only predictable future is the *unpredictable*, and the unexpected should be prepared for.
- logistic trade-offs have to be more carefully weighed in terms of economies of scale of production and delivery lead times.
- local warehouses and distribution points have been reduced for reasons of rationalisation. This incurs penalties in terms of delivery speed and cost.
- its regulatory constraints are increasing (not only pharmaceutical and financial but also ecological).
- in order to have overall knowledge of product and value flow*, it is necessary to have a centralized information flow.

* Value Flow see end of chapter p. XXIII.

- it is beginning to see the vital necessity for managers all along the chain to grasp, understand, and calculate *the value from one end of the pipeline to another i.e. not only the mechanics, but the mathematics and then the economics of the pipeline.*
- it is learning that it must know the administrative and other costs related to chain interfaces i.e. the details of the cost of the Supply Chain.
- it must do Benchmarking outside its area of business in order to continue to improve and embrace agile business techniques.
- it has realised that it is always necessary to challenge inappropriate lead time and constraints which have been amassed over the years both by staff and computer systems.
- the realisation that near instantaneous information on order movement, or differences, are communicated all along the cycle *.
- lastly and perhaps most importantly, it is realising that trust and loyalty are vital managerial tools about which very little is known. Without trust a supply chain cannot function; with loyalty it becomes a powerful competitive advantage. Trust and Loyalty must be understood in all its cultural, psychological, and rational dimensions before it can be practised effectively.

It is in consequence of all the above, that the industry is realising that the Supply Chain is an important strategic issue where top management commitment and drive are essential for developing its business.

One of the problems in establishing new rules, guidelines, or practices resulting from new experiences, and best practices, is that there is generally no adequate framework. This is due to the

---

* This is known as the "order penetration point" meeting the pre-established plan i.e. an order, or a set of orders, has to meet a predetermined trigger point (quantity) which sets the reordering and remaking plan in motion.

This is supposedly based on "economic order quantities" – which is generally too late! Hence the realisation of the need for warning systems.

fact that the overall vision of the practices and its aims are not clearly seen in an holistic fashion.

Furthermore, Supply Chain is a multifaceted subject; multifaceted because it involves suppliers, manufacturers, users, commercial, and regulatory, as well as logistical, and of course financial issues.

Therefore, there is a need for a framework in order to create a basic "transfunctional" understanding of the supply chain and the problems and opportunities in its practice.

The Guide will hopefully facilitate the understanding of the Supply Chain.

It will permit more rapid and precise anticipation of possible future events.

It will enable a more effective response to changing situations.

It will enable one to control and manage better the actual cost of the supply chain.

This will enable partnerships and contracting to be effected in both a more meaningful and business-like manner.

It will enable one to assess risk in a more global and precise way.

The final client, as well as each intermediate client, will be better satisfied.

It will facilitate and provide a logic to carry out audits and draw up quality specifications.

It will enable the establishment of continuous improvement programmes, with co-ordonated and aligned objectives.

It will help in the understanding of the breaking down of barriers between individuals, between departments inside a company and barriers between a company and its external world, giving benefits to all concerned.

Quality and Service are the key to successful Pharmaceutical Business.

The Supply Chain, with its primary aim to reduce costs and increase efficiency and productivity, must at all times keep quality and service in focus for products prior, during, and after manufacture and supply. *This means having the right product at the right time and in the right place.*

As the whole system is already under pressure, any disfunctioning would add further pressure to one or more links in the chain. This would endanger quality by cutting corners, taking excessive risks, and substituting from other non-fully proven sources. As in a busy motorway where cars move very fast, and there is a problem with one car, it could engender a multiple crash situation – the snow ball effect.

This Guide is here to help to focus on the prime aims of Quality and Service.

The main themes of the Supply Chain are:
– Improving Reliability,
– Logistics,
– Supplier Relations,
– Manufacturing Practices,
– Manufacturing Efficiencies,
– Systems and their standardisation,
– Human, Cultural, Managerial, and Teaching,
– Human Relationships,
– Commercial
– Regulatory
– Distribution methods
– Distribution channels
– Distribution distortions
– Quality and Quality Assurance are interwoven at all stages,
– Linkages.
– Measures

The successful practice of the Supply Chain depends, among other things:
  1) Identification of key critical points
  2) Obviating of key critical points
  3) Every link understanding every other link,
  4) Understanding the administrative cost of the chain,
  5) Clarity and understanding of definitions,
  6) Instantaneous and agile information flow,
  7) Communicating,
  8) Early warning systems,

9) Trust, Loyalty,
12) Risk assessment and sharing,
13) Responsibility sharing,
14) Learning of new functions,
15) Learning of new outlook,
16) Continuously improving.

The key quality aspects comprise of:
– Clear specifications,
– Pin pointing weak or potential weak points,,
– Avoiding or circumventing weak points,
– Reinforcing weak points,
– Fast feed back systems,
– Audits,
– Scanning systems,
  (early warning systems and feed back for reacting to them),
– Knowledge sharing reflex,
– Clear specifications,
– Partnering.

These are the themes which will be discussed in this book, with as many examples as possible, based on real experience.

Value Flow (from page XIX)

Value Flow is a new concept where it is more important to know from a financial point of view how the values of a company's product flow. And therefore where these values are. Also where the different values are added, rather than how many boxes were made. The products have no final value until the last stage of the process and its last sales value (price) is known.

Value Flow includes knowledge of all parts of the Supply Chain. It takes into account trade-offs between volumes and service/availability factors, as well as the cost of the pipeline and inventory costs.

Loss of value is also incurred by reduced shelf life for finished pharmaceutical products.

If the NPV (Net Present Value) concept were applied to manufactured goods and the value of work in progress thus assessed, the actual cost of goods would be considerably higher than the way cost of goods are calculated in the classical manner.

This cost or loss in value, is in fact incurred, and reduces a company's margin in an unseen manner. Like a ghost – the non added value ghost eats up some of the profits!

"We seek him here, we seek him there
Those accountants seek him everywhere
Is he work in progress or stock in hell
That damned elusive wasterell".
(With apologies to Baroness Orzcy and the Scarlet Pimpernel).

Value Flow underlines the difference between value adding and cost adding such as is induced by the double handling of goods, or unnecessary retesting in different areas. In the latter case, time has elapsed which has also added cost, whilst the value of the product, instead of increasing, has decreased due to cost incurrence.

Value Flow concentrates vision on Lead Time as a whole.

Thus interfaces between segments or components of the lead time can be more closely examined in order to reduce or eliminate them.

1 ISSUES, SCOPE, HISTORICAL PERSPECTIVES AND PRINCIPLES

SUPPLY CHAIN ISSUES – WHY'S – HOW TO AVOID OR REMEDY THEM

The Supply Chain is all about shortening the cycle between a number of source points and the user. This implies a massive increase in speed both of physical and administrative transactions.

Agility, speed, and reactivity, plus new business and technical processes, induce quality issues and risk issues which previously have been non existent, hidden, or reduced by the slack and buffers in the systems.

The critical issues have to be identified and means found to avoid them. Also, systems must be put in to monitor them. A line of sight scanning system is proposed, which is an early warning system with a feed back process to ensure corrective action.

The risks associated with these issues have to be more carefully evaluated and clearly known by all along the chain. Many of these risks are shared, thus these responsibilities have to be known. It is based on sharing culture between partners.

Agile Manufacturing and other agile business techniques:

Agile manufacturing means not only anticipation and fast action to unexpected events. It also means having simultaneous options for different outcomes. It means having parallel and alternative action plans. It means reducing speed suddenly: even stopping and sometimes going into reverse.

But it means all this without incurring penalties, or at least without incurring *unassessed* penalties.

This is the only way that change can be mastered in an economic manner.

The Supply Chain is one of the key pillars of Agile

Manufacturing. The others being I.T. (Information Technology), multiskilling, holistic business vision, rapid information gathering, structured action planning, integrated cost accounting, real time, and accurate market feed back, etc...

The mental attitude of staff is vital in agile manufacture. Not only lively, curious, and decisive minds are required. But also a capacity to handle contradictions, dilemnas, multiple choices, rapid mental modelling, and above all an holistic understanding of the business, combined with an open mind is necessary.

This subject is discussed in more detail on page 28 and onward; see Unpredictability and Anticipation in Agile Manufacturing, in Chapter 2.

It should perhaps be clarified that a key feature of the holistic business view is a total cost view on all decisions and evaluation of options. By total cost, we mean the assessment of individual costs and weighing the total of these on the impact of the whole business. Parochial cost evaluation could seriously jeopardise total business efficiency. An example could be the strict adherence to a departmental budget avoiding an investment for a machine which could also be used by another department. Avoiding the use of unbudgeted overtime needed to satisfy an order is another typical example of parochial and egoistic budgetary thinking. Agile accounting means just this, thinking not only in strict budgetary terms but aligning rapidly value with total business interest.

Definition and Scope:

DEFINITION OF THE SUPPLY CHAIN:

The objectives are that the flow of all that goes in and around a pharmaceutical product, from the suppliers of components to the finished product and services, is delivered to the customer in the most economic and timely manner, and at the quality that the user expects.

Scope:

Its scope includes:

Product:	– Design	plus specifications
	– Process	» »
	– Packaging	» »
	– Quality	» »
	– Manufacturability	
	– Customer satisfaction	

Logistics:
- Forecasting
- Resource planning
- Scheduling
- Material handling
- Storing
- Transport

Suppliers:
- Contracts
- Specifications
- Flexibility
- Information Flow
- Two way commitment
- Monitoring
- Audits
- Training

Information Technology:
- Strategy
- Systems
- Resource priorities

Plant utilisation:
- Rationalisation
- Capacities
- Shift and working hours
- Optimising large and small scale throughputs

Regulatory:
- Submission data
- Compliance
- Validation:
  - equipment,
  - batch sizes,
  - specifications

- SUPAC
- Free Sale Certificate
- G.M.P. & N.D.A inspections

New marketing channels:
- Pharmacies
- Licensing
- Alliances
- Emerging country problems

Measures:
(Metrics)
- Customer Oriented
- Meaningful
- Accessible
- Consistent
- Benchmarking
- Continuous improvement
- Service levels
- Quality
- Cost
- Responsiveness
- Flexibility

Communication:
- Feed back from Market to all levels (fast)
- objectives to be aligned, coherent, prioritised and customer oriented
within Company, between Departments between Functions, with Suppliers and Distributors
- Objectives to Understand (Educate)
Empower
Obtain Commitment and Ownership

Sourcing:
- Global, regional, local
- Parallel imports and distortions
- Outsourcing

Quality: – Objectives integrated throughout Chain

Cultural: – Customer response behaviour
                     – Mind Sets, Vision of Chain
                     – Transparency, Trust, Loyalty
                     – Risks

Customer: – Identification of different customers
                     – Anticipate needs
                     – Define needs
                     – Satisfaction levels
                     – Negotiations

The scope therefore covers everything from the design through to procurement, manufacture, dispatch, delivery and payment of medicines at the right cost, time, place, and quality. Its scope is much larger than manufacturing, which is just one part of the chain. Human Relations play a vital role throughout the chain.

Its aim is economy, compatible with the above mentioned criteria.

The financial benefit realised is the bottom line. The gains go straight to the profit line!

Viewed from another aspect in the manufacturing field, the economies can best be achieved by eliminating waste.

Waste of: – excess stock   raw materials,
                                    work in progress,
                                    finished goods.

             – rework, rejects,
             – time   procuring
                                    manufacturing,
                                    delivery,
                                    materials handling.

             – administrative work,
             – interfaces,
             – circuits,
             – space,
             – retesting,

- energy,
- capacity,
- materials,
- speed in new product introductions,
- effort by people

– It means integrating forecasts, schedules, planning into preferably one, or at least into a closely interlinked, system.

– It means professional negotiations.

– It means ordering and delivering

- at the right place,
- at the right frequency

making

- the right quality,
- the right quantity.

– It means adding value at the right place and time i.e. in the most economical manner.

THE STORY OF MANUFACTURING IN THE PHARMACEUTICAL INDUSTRY FROM MANUFACTURING PUSH TO CUSTOMER PULL. THE BIRTH OF THE SUPPLY CHAIN – AN HISTORICAL PERSPECTIVE (See Figure 1.1.)

In the past, until the 1960's, manufacturing was PUSHED by technical people to get the maximum product out, despite difficulties of imperfect production equipment and poorly understood processes.

Then eventually these problems were resolved, processes became better, more reliable, and often simpler. Machines improved with experience, with more professional manufacturers, and also better maintenance and up-keep.

Manufacturers sold and pushed what their research found.

So products flowed through easier and stocks became higher. The cost of the product was not excessive and margins were high.

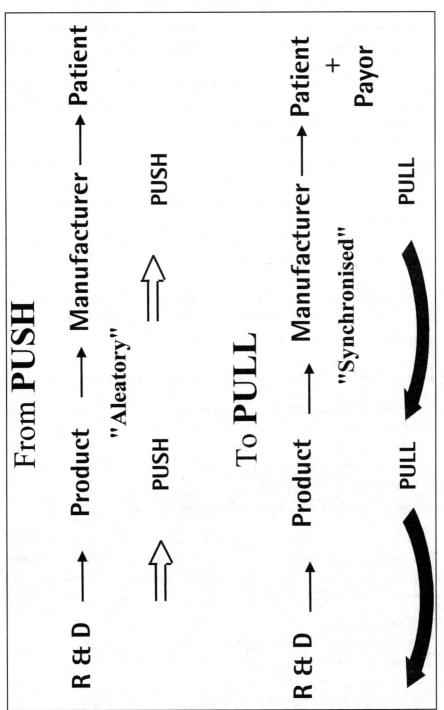

Figure 1.1

This was the period when Good Manufacturing Practices and Quality Assurance came to the fore as a result of a number of dramatic accidents.

Costs started to rise notably after the 1973 petrol crisis when energy became 10 times more expensive. Wages started to rise, and inflation ran at double figures.

Productivity and Efficiency came to the fore. In looking for Quality problem solving, notably in Japan, other productivity improvement techniques were found. (The fact that some of these techniques came originally from the U.S. and were "lost" in the "roaring sixties" due to other priorities, proves the story.)

Just-In-Time, Kanban, and other Japanese productivity tools and measures were introduced, studied, and adapted, thus Lean Production came about.

Subsequently – (notably after 1992), with rationalisations, mergers, and the realisation of the cost of large stocks, manufacturing started to be PULLED.

Manufacturers researched and made what the patient wanted or needed!

Re-engineering, that is looking at manufacture from a process perspective, made it necessary to look at the Supply Chain. Firstly, for its financial potentials and then in terms of systems and philosophy and lead time reduction.

Manufacturers made what payors were prepared to pay for.

Since then the pressure has mounted because of the need to lower costs and increase margins. The supply chain is at the moment the ideal Zoom Ocular (Tele and Microscope) to examine (See Figure 1.2):

a) the process,
b) the supplier relationships,
c) the transport arrangements,
d) stocks and volumes levels,
e) speed and lead times.

Supply Chain–ideal optic to view :

- **Process**
- **Supplier relations**
- **Stocks and volumes, Transport**

and one discovered

Insufficient Staff Training
Poor Maintenance

High Transaction Costs
The cost of Time

Inappropriate Factory Organisation

Figure 1.2

By looking at these, further inefficiencies have been discovered in terms of:

- Transaction costs,
- Hierarchical levels,
- Staff training,
- Decision making processes,
- The cost of time,
- Maintenance methods,
- Factory structure.

It was realised that as the factories become more complex, serving many markets, and having many clients, having more different packs – the staff, including many managers, did not fully understand the Supply Process or the Supply Chain.

Figure 1.3 sums up this historical review.

The industry started to benchmark outside its own field. Agile manufacturing and agile business techniques were started to be practised, where the need to react quickly to unexpected events were realised. This required new capabilities, new mind sets, an ability to anticipate and practice foresight.

The supply chain is key in agile business techniques, because it includes the "extended" supply chain, that is all suppliers and clients.

A chain goes for from A to Z , any single weak or broken link weakens or breaks the chain.

No matter how fast and efficiently a tablet press works, if the tubes into which they go are not clean, or are not there at the end of the line – there is no product – or there is large waste of time.

Many major commercial issues such as transfer prices, parallel trade, export quotas, import licences, wholesaler and trading arrangements are intimately connected to the Supply Chain. The same optics can be used in order to understand and relate them better to the overall pharmaceutical business.

I believe that the present way of looking at, understanding, acting upon, and improving the manufacturing process, is through the Optics of the Supply Chain. This leads us into

DEVELOPMENT OF MANUFACTURING FACTORS OF TRENDING CHARACTERISTICS

PERIOD	MAJOR ISSUES CHALLENGE	CHARACTERISTICS	MARGINS	PRICE
Pre 50's	Manufacturing Techniques and Process Packing techniques (As output increased mistakes happened) R & D	R & D Push Production Objectives : Large Labour Force Push Manufacturing, No computers, Low Labour costs, Low Indirect costs.	Low	Low
60's		Training of staff (difficulty finding right staff) Improving machines	Medium	Medium
70's	GMP	Building Factories " large numbers Labour costs going up Indirect Labour going up Regulatory + Low level use of computer	High	
80's	World Class Manufacturing Cult of Excellence Pull Manuf JIT	Modernisation of Factories, Mega products, Automation Massive computerisation & Training Programmes Regulatory ++ Lean Manufacturing Lowering of Labour utilisation Soaring indirect cost	High	Very High
90's	Reengineering Supply Chain Agile Manufacturing	Further Automation Large Indirect Labour Costs Complexity, Globalisation Regionalisation Patient Pull - Anticipatory Techniques	Low	High
2000	Medicine Value Management + Mass Customization	Personalised Industrial Production Regulatory +++ P.I.P.	Medium	Lower

Figure 1.3

organisation or reorganisation and a better understanding of re-engineering and put us on the road to agile manufacturing and agile business management. This in turn will eventually lead into Mass Customisation.

I feel that this story requires spelling out, firstly because many talk about the Supply Chain : Re-engineering, Process design and redesign without fully understanding the why. Secondly, many people looking at the systems and ways of working, do not understand how and why they originated in the way they did.

Agile Business Techniques and the Supply Chain

Agile business techniques are a set of practical concepts which help companies to *respond competitively* to *customer demands*, which are *evolving* and changing fast, in an *economic* and *social environment*, which is *also changing fast*.

The characteristics of these changes are that they are *unpredictable*, and therefore cannot be accurately planned for. They can however, by a series of techniques, be *anticipated*. By this means, one can react efficiently.

For the next ten years, agile business techniques are likely to be key management precepts.

Anticipatory techniques consist, among other things, of:
– Active 360° intelligence,
– Imaginative outcome scenarios,
– Having simultaneous options,
– Alternative action plans,
– Energetic and systematic follow up of all decisions and actions,
– Instantaneous information flow (see Agile Information Flow, page 178),
– Reactivity culture
– Risk taking.

Other characteristics of agile businesses are:

– A cross functional reach whereby one can act, inform, and take decisions in a process view, or total task oriented manner.

– Internal and external consulting and decision taking avoids, to a maximum, the hierarchical circuits.

– Empowerment is largely practiced.

– Administrative procedures are reduced to a minimum in order to save time and cost, this is replaced by a strong Trust and Loyalty culture.

– Managers are trained to prioritise among conflicting objectives, and to deal with dilemnas.

– Flexibility, transparency, and open minds, are other cultural qualities which are required in an agile business environment.

– Use of technologies enabling fast adaptations.

The supply chain is one of the key competitive elements in agile manufacturing.

Establishing partnership and transparent business dealings with suppliers, using instantaneous information exchange, trust, and loyalty are being used to reduce heavy contractual and administrative procedures.

The actual manufacturing and packing processes should be aimed at short cycle times, rapid machine change-overs, and the use of staff who are multiskilled and who are able and willing to work flexible times (including week-ends and night shifts) when required. In other words short lead times.

Spare capacity, or means to create capacity quickly in terms of equipment, machines, and space, should exist without penalising costs. This is vital in the pharmaceutical industry where loss of sales due to out-of-stock situations might be far more costly than having the spare capacity.

Marketing wishes and desiderata must be aligned with technical capabilities, in other words, technical development must be in line with marketing requirements both in terms of time scale and feasability.

Maintenance and repair procedures, in order to be agile, should be reactive. Preventive maintenance and close monitoring of machine and equipment performance are vital. Follow up and project control are essential to keep to tight schedules.

Agile transport and shipping arrangements are based on the judicious use of carefully planned alternatives and opportunistic rapid change from one means of transport to another, or from the use of one firm to the use of another.

Contact with marketing, having instantaneous sales information and commercial feed-back, permit agile reaction both at the supply and distribution level.

Information flow is of course the key to all these matters, see page 178.

A knowledge of total cost, irrespective of individual functional cost structures and overhead recovery basis, is absolutely essential in an agile business. Only thus can the right decisions be arrived at rapidly, and fairly accurately. Precise figures are not absolutely essential – rapidity and rough cut precision outweigh the time lost in precise calculations, which are in any case mostly inaccurate. Activity Based Costing is obviously a preferred way of obtaining figures.

Information Technology is helping the better and faster use of information from the supplier to manufacturer, and from marketing backwards through the chain. Faster information on equipment use and process efficiency help to reduce cycle time and reduce waste. Better programming is made possible by the combined use of the above information.

Agile client intelligence tries to find out what the client wants, what he needs, and how best and most quickly and economically he can be not only be satisfied, but made so contented with the product and service that he will always want to come back. Client Satisfaction Monitoring systems are part of the Information Technology system.

The same principles apply to product development, pack changes, and Regulatory and Quality Assurance matters. Many design and regulatory problems can be avoided by a more anticipatory approach and the pursuing of a number of parallel options in order to choose the most adequate one at the appropriate time. Similarly, many projects could be stopped earlier in order to pursue others.

A framework encompassing some of the above concepts is shown below. Figure 1.4.

Agility is not a short term management precept. It is a philosophy which never accepts "No" for an answer, where difficulties and barriers are overcome, not necessarily by financial means, but by will, perseverance, and an attitude.

Alternatively, difficulties are overcome by looking at problems in a different and innovative way. The cartesian thinking and reasoning does not generally give agile answers.

In agile business technique management, the high exigencies which are required in the Pharmaceutical Industry must at all times be exercised. Due to the speed of events, an acute awareness of quality issues has to exist. *Exigence* means demanding and has a second meaning – urgent. Agile manufacture in the pharmaceutical business has "exigence" as its principal and double characteristic.

Agile business techniques are not easy to introduce into pharmaceutical operations. Many excuses for not doing things, for not going so fast are put forward. This is generally on the grounds of quality and risks associated with it, as well as the heaviness of the regulatory process.

The main problem is the "compliant culture"* which is generally engendered by fear of the consequences of mistakes, and to blind adherence to procedures and regulatory requirements. This induces heavy administrative working methods, where belts and braces as well as parachutes are used to protect oneself from reproaches or criticism. Scientific and cartesian cultures do not lend themselves to agile thinking and action.

Whilst these barriers can exist, there are however means of surmounting or avoiding them. The past habits and mind sets are the most difficult to change and transform into a fast speed dynamic. *Acceleration requires much more energy than sustaining high speed !*

* see ANNEX 2: Culture Changes necessary in the Pharmaceutical Industry.

A FRAMEWORK FOR
AGILE BUSINESS AND THE SUPPLY CHAIN
(in the Pharmaceutical Industry)

AGILE TECHNIQUES	AGILE MANAGERIAL SKILL	FIELDS OF APPLICATION
STAFF : Multiskilling Flexible working methods	Anticipation Foresight Alternative Action Planning	Suppliers Partners
FINANCE : Cross Functional Accounting Total Cost Evaluation	Simultaneous option evaluation Instantaneous Information Flow « « Reception	Development Regulatory
ADMINISTRATION : Minimum Procedures	Managing Conflicting Objectives « Dilemnas Prioritisation and Rapid Reprioritisation	Quality Control / Assurance Manufacturing
INFORMATION TECHNOLOGY : Integrated Better programming Better Process Control	Cross Functional Reach Non Hierarchical Contact Open Mind Holistic view (non functional, long term) Transparency	Packing
TECHNICAL : Intelligence Flexible Capability « Capacity Alternative equipment « machinery	Change Culture Empowerment, Delegation Risk Sharing Trust Creativity Loyalty	Maintenance / Engineering Storage Dispatch
LOGISTICS : Alternative sourcing Partnership negociations Alternative delivery methods Outsourcing Insourcing capability	Cross Industry Benchmarking and Competence Webbing Non administrative behaviour Curiosity	Transport Project Engineering
CLIENT : Intelligence Rapid Feed back Satisfaction Criteria Monitoring	Barrier breaking Identify with internal and external CUSTOMER Rapid Reactive	Projects

Similarly the start-up into agility requires a lot of effort and push. However, once the results are seen and seen more quickly than in the past, then it is easier to maintain. The automatic and energetic follow up of actions and decisions is a reflex which people get accustomed to, and must not be interpreted as a lack of trust.

Agile Engineering

Simultaneous engineering concerns the subject of the building of factories, without knowing exactly what products could be manufactured in them. It refers to ordering and designing equipment without knowing the total process cycle. This obviously involves risk-taking and some guess work, but the time saved compensates for the errors and slight deviations which might have been made.

Obviously experience, judgement, and the courage to take decisions are necessary. Whilst engineering is not the subject of this book, the making available of capacity is. Rapid and flexible building, construction, and adapting methods give big competitive advantages both in obtaining new work and in confronting sudden and unexpected increases in volume.

Agile Decisions Making

The agile decision making process has some features which require to be explained.

Firstly, it is a rapid process. Secondly it must be clear in order that it should be understood by everyone unequivocally and must not be fudged. Thirdly, it should up to a point in time be reversible whenever possible, so that another decision in its place can be made. This may seem contradictory, however, agile management is based on swiftness, alternative options, multiple action plans and flexibility and the ability to live with contradiction. An example related to the supply chain may be useful.

A decision to launch a new pack, as fast as possible, has been taken, the design, artwork, weight of paper for the carton, and method of closure have all been pushed through, by teams, in a very short space of time. The carton was ordered to be delivered in 7 days time. Suddenly, marketing intelligence finds out that for tactical reasons, it would be better to launch 3 months later.

In the agile management mode, this information must be communicated *within the hour* to the carton manufacturer in order to see at what stage the order is, whether the paper has been ordered ; whether it has been printed or cut, etc...

This way there is the maximum chance to halt the work in progress, which could be continued later. This way, expensive making and storing could be obviated. The key in this example is the speed of doing something in the first place and the speed of stopping, or at least slowing down the same thing with alacrity.

Another example concerns the purchase of a machine. Two factories both had and unforeseen increase in sales of the same product. Both had a lack of cartoning machine. Factory A phoned around the machine suppliers and found one available. It was immediately ordered by fax. The machine was operational 7 days later.

Factory B decided that it had to undertake a comparative study between different makes and models – this took three weeks. Eventualy it arrived at the conclusion that the best machine was the one ordered by factory A. The order was passed through the normal channels but marked urgent. The order took two weeks to process (instead of the usual four). Therefore five weeks were lost by Factory B. The lead time of the machine was four months. Therefore the machine arrived in factory B six months after the machine ordered by Factory A received theirs.

Another example concerns the installation of a small sterile facility. In factory A (which used movable partitions) the sterile room was installed in five weeks. In factory B, which has solid wall construction and who insisted that sterile rooms could not be made with movable partitions it took ten weeks to install a similar facility as that installed in factory A. The agility in this example is not only due to the use of movable partitions, but also the whole approach to project management, where long drawn out contracts and very precise quotations are used. Precise quotations take a long time to elaborate and are not always necessary.

Examples in building factories, validating equipment, training staff, and in approving suppliers abound. But here we are perhaps deviating from the subject of this book!

Agile Team Work

The best example generally taken to show agile team work is the game of rugby. In this game, the ball is passed from one player to another to avoid falling in the hands of the opponent and to make sure that the team can score.

Six things are essential in this team work:
1) each player knows his role,
2)   »   »   the role of the other players,
3) each player knows how to interchange for another player,
4) each player plays for the team and not for himself and his personal advantage,
5) each player plays for the aim of the game i.e. scoring and winning,
6) each player knows the rules of the game and the size and shape of the goal posts and their position,
7) training and practice.

All these qualities are needed and apply in agile management practice with the exception of item 6, because in business some rules change others don't; the size and shape of the goal post might also change.

These unpredictables are nowhere as foreseeable as the change in tactics which an opposite rugby team may employ. Generally whilst these tactics may change during the game, they are well known and still fall within the "rules" of the game.

Team work is very useful for the 360° Active Intelligence. It is also very useful in working out alternative options, and in making simultaneous action plans. When it comes to decisions, unless the team is both geographically and mentally very close, the team is not generally sufficiently agile. Therefore an individual must be made responsible for agile decision taking.

In other words, teams (especially large ones) are generally too slow in taking rapid decisions. There are, however, exceptions which prove the rule. The author has known tightly knit, close working people who knew not only each others minds and the way they were working, but were also constantly made aware of changing surroundings and requirements. The result was that team decisions could be thus be taken without a lot of consultation and hence loss of time.

Agile Techniques and Learning Organisation

This book often refers to the Learning Organisation and its merits. How does the learning organisation fit in with agile management ? The answer is very simple – the learning has to be done in an agile manner. This means one has to learn:

1) fast,
2) all the time,
3) to change the subject or content of learning without hesitation,
4) not to hesitate to unlearn, that is to change the way one has learned to do something. Learning fast is a habit which most people can be trained for. Often fast evolving situations create the pressure to learn fast, in which case the unlearning – relearning process is simple.

GENERAL PRINCIPLES OF QUALITY ISSUES IN GOOD SUPPLY CHAIN PRACTICES

In the following chapters of this book, the main themes of the supply chain are discussed. The principles of good practices are set out, the reasons for these are given, and examples are used to illustrate some of the points. Figure 1.5 outlines the supply chain in the industry.

An overview of the main quality issues are listed below:
– Identification of Critical Quality factors,
– Enumeration of Quality Actions,
– Establishing Critical Quality Indicators,
– Recording of factors from which fast decisions can be taken,
– Establishing relevant Macro (overall sales) and Micro (local or regional) trends,
– Having interactive database for: Demand, Stock, Cycles, and Lead times.
– Establishing "flashing indicators" for tracking purposes,
– Having an Information Technology system to scan flashing indicators,
– Maintaining traceability of products and their components,
– Monitoring new weaknesses in system,
– Inculcating Trust and Loyalty,
– Training staff to have total line of sight or vision of cycle, especially relative to importance of chains and length of cycle,
– Communicating with rapidity, transparency and clarity.
– Risk assessment and the sharing of responsibility.

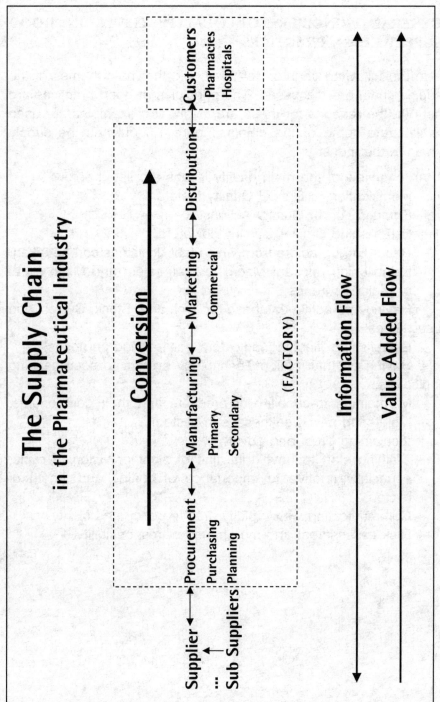

Figure 1.5

SOME OTHER PRINCIPLES RELATED TO THE SUPPLY CHAIN:

Some general principles held by the author are very relevant to the supply chain. They are as follows.

- A supplier, whether internal or external, should be treated on an equal footing in any dealing. And therefore should be considered as a Partner (the term supplier implies a subservient or secondary role).
- Each partner should make every effort to understand the need of the other. It is an inacceptable stance to say that the customer did not know what he wants. The partner or supplier must help him to express his needs. As a corrollary the supplier is bound to offer the best, and/or the most effective product or service. This is the real meaning of Customer Services.
- For the chain to work optimally, it is essential to have a fully **Informed, Committed, Involved,** and **Motivated** staff. The staff must **Identify** itself clearly with the company, its mission, and the needs of its **Clients**.
- In order that staff should be able to identify, it is of great advantage to have maximum responsibility on a local or at least regional basis. Any centralised control activity is a barrier to the "feel" of the market, customer, and indeed local suppliers. In this case, great effort has to be dispensed to endeavour to overcome the barrier. This effort could be more usefully deployed elsewhere.
- Short term objectives and compartmentalised budgetary thinking are major obstacles to the proper understanding and execution of the supply chain. A holistic mind set, focussing on the business as a whole, is essential in all decisions and in determining priorities.
- Information must be transmitted clearly and promptly. Very often "bad" news is communicated too late for adequate action to be taken. This is perhaps out of fear, or the hope that the news will turn out to be not bad.

- Some communication can only be effective if it is verbal. Some, if it is face to face. Over reliance on writing only, is dangerous.
- Great and sustained effort must be placed on breaking down corporatist, functional, and other internal barriers which automatically hinder the supply chain.

- The more foresight and anticipation, regarding customer requirements and suppliers' capabilities is used – the more effective and agile the supply chain becomes.

- Common sense should always prevail over administrative procedures. The "rule-compliance – procedure" culture must be attenuated by managerial responsibility. See Annex 2 on Culture Changes necessary in the Pharmaceutical Industry.

- Holistic management means operating within the "whole" and seeing the "whole" in each of its parts. But (to the author), it is also the ability to use simultaneously different management modes, ranging from autocratic to full empowering, depending on the needs of the business, on circumstances, and on people. This necessarily requires a culture which accepts the coherence of apparently contradictory ways of managing. A lot of time, learning, and discipline is required to obtain the stretch and maintain the spur for initiative, creativity, and entrepreneurship, within this Holomodic framework.

- Competitive supply chains require good systems, (which others may have too), but they require above all, *trained* managers who *inspire* and *enable* their staff to *perform* to *extraordinary standards* in a *sustained* manner. Team work and empowerment should not be a substitute for a sense of urgency, decisiveness, firmness and accountability. Each individual's ideas, understanding and commitment make the supply chain a success.

2 FORECASTING

Forecasting is a crucial activity for both operational and strategic reasons.

The commercial vision, which is driving forecasts, is necessary for the planning of capacities, and the reactivity and agility to product sales fluctuations.

The accuracy and reliability of the forecast at both the strategic and operational level (i.e. long term and short term) is, of course, the basis for optimising and therefore shortening cycle times.

This, naturally, is not always easy. What however is easier, is to ensure immediate feedback of market trends and sales forecasts. The decision for creating and changing forecasts has also to be either shared or clearly taken.

The sharing of responsibilities is a key feature running throughout modern supply chain management.

The changing, or modifications to forecasts, have to be done rapidly and decisively.

Often, this is fudged*. Similarly, if the decision to modify has been taken, it must be known throughout the organisation.

Often marketing, or whoever is responsible for forecasting, think that the only people interested are the factory staff. Whereas in reality, many suppliers upstream are concerned, who themselves have much longer cycle times and lead times. Furthermore, critical suppliers who have bottle-necks themselves, quality problems or potential material shortages, have to be identified and closely monitored. (see Fig. 2.1, showing the

---

* It is not a crime to see no further than possible at a certain point in time. It is normal that as time goes by, one's perspective extends further, – unexpected things happen. The importance is to *realise change*, or new data *quickly*, and *react to them rapidly*.

supply chain of a finished pack of aerosol. In this case, there are about 25 components which make up the final pack. Many of these components have sub components. All of these have to be forecast, made, stored, tested, etc...).

Similarly, marketing is influenced more by short term sales and by the financial objectives set by their boss. They do not realise all the ramifications on the supply chain.

Sometimes even financial people influence, or try to influence forecasts, and of course investments, by calculating minimum return on investment quantities.

A far as short term forecasts are concerned, ideally, there should be an interactive database. Thus when a pharmacist (or whoever) dispenses a box of medicine he reads the bar code and this information is simultaneously relayed back to:

- the Distributor,
- the Warehouse,
- the Manufacture,
- the Purchasing Department,
- the Component Manufacturers.

Although this is a simplistic model, it is an ideal towards which one should strive. Certainly parts of this information system can be easily achieved – let's say from the pharmacist to the manufacturer or purchasing department. Beyond this a lot of collation, deduction and decision making processes intervene.

These are difficult or impossible to automate entirely. Also from this point, a lot of other suppliers are involved to different degrees of importance and urgency.

However, the human intervention process must be *lightened and shortened* – this is the key to successful supply chain Management.

It is necessary to evaluate and always to know the risk attached to the precision of the forecast.

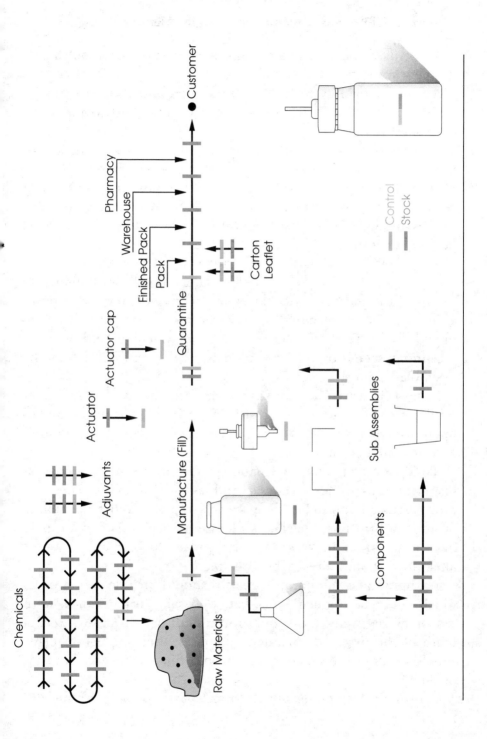

Unpredictability and Anticipation in Agile Manufacturing:

The only certainty is change, this adage is now also applying to the pharmaceutical industry.

A firm's capability to foresee, anticipate, and adapt to change rapidly and effectively, shows its strength and competitive advantage.

The mind set to accommodate the apparent contradiction of unpredictability and anticipation is very special. It requires not necessarily intellectual brilliance or a schizophrenic mind, but a certain flexibility and a wide, holistic view. A thorough understanding of the possible impact of events on the process, people and capacity is necessary.

This subject is of particular pertinence in forecasting, for the reasons mentioned above, as well as later on in this chapter.

Agile manufacture and business techniques revolve around anticipating and reacting. How can one be prepared for the unexpected? Foresight and perception can be cultivated by imagining scenarios, making models, and creative thinking by assembled minds.

Logical i.e. cartesian thinking and reasoning, *as well* as lateral thinking and intuitive modes should be used. It is important to build a climate of confidence where "crazy" ideas are permitted and not laughed at.

Rapid, simultaneous information gathering from a number of disparate sources are scanned and assessed rapidly. They are then discussed and debated, again very rapidly ; and a decision is taken in a firm manner. The scenarios or decisions chosen must have a wayback, in case they are proven wrong and another alternative should already be available.

Maximum and minimum scenarios should include action plans which could be applied immediately. If sales, for instance, were ten or twenty times more than forecasted; or conversely, were a tenth of the originally planned quantities, then the action plans should minimise the consequences both financially, and in terms of time.

Curiosity in searching out historical parallels, both within and outside one's business, a wider understanding of possible impacts from marketing and a commercial, as well as technical aspects, is required. This will allow more rapid anticipation which is very relevant in new launches, relaunches, product withdrawals (and a variety of other subjects, not only related to forecasting).

This should be done on an informal basis, as often as possible or necessary; and should become second nature in a firm and part of culture.

Accuracy, reliability and knowing precision:

Reliability is more significant than accuracy. For, if the degree of accuracy is known, i.e. something which we know may vary between ± 20 %, one can plan production and suppliers accordingly. What, however, is more important is that this variation be **reliable.** That in our case, there are no variations of plus 50 % or minus 75 %. Reliability also implies that if a change in forecast is anticipated, it is immediately communicated along the chain.

In case of a new product launch, the criteria and sensitivities are different and these are discussed in more detail later on in this chapter.

Long term forecasts:

Long term strategic forecasts are necessary for capacity and capability investment.

By capacity we mean the quantities to be produced over a certain time, – a rate of production.

By capability, we mean not only a variability in output but also the ability to change a machine or equipment from one pack size or batch size to another, in a given time. It is the flexibility which is required, either for the type of product, the type of market, or both.

It is not often realised that some products when they reach a certain volume of throughput, might require:

- new factories or factory extensions,
- new departments,
- new equipment,
- new air conditioning and other services,
- new storage space,
- new staff,
- new training programmes.

These are all within the manufacturing organisation and require both time and money – usually in copious amounts. A new factory could take 2 or 3 years before it comes on stream and could cost anything from $ 15 million to $ 50 million or even more.

Equipment nowadays could take a year or more to be delivered. This is without taking account of the time required for the specifying, ordering, and commissioning of the equipment. Validation might take three months, and in the case of new technology, could take up to eighteen months.

Now all these "new" items imply quality issues – quality directly related to the product.

When sales increase, leading to further investment in equipment or space, one generally buys different, larger, and more modern equipment to replace the existing ones. This equipment has to be qualified, validated, and often approved by regulatory authorities.

Often a process is changed, generally with the intention of improving either quality, efficiency, or preferably both.

A company once increased its sales of caragheneen extract tablets by a factor of 3. The classical slow process of granulation in mixers and drying in ovens was transferred to fluid bed granulating and drying equipment.

This improved efficiency by a factor of 10. Cycle times were reduced from 15 hours to 2. But the whole process had to be revalidated and reregistered.

The flow properties of the granules had become different and even the taste changed. (The peppermint flavour was less affected by the fluid bed drying than by oven drying).

When one is running near to an out of stock situation because of lack of capacity, the reduction by 13 hours in lead time counts a great deal!

Similarly, more recent examples are to be found with the changing of the ranitidine tableting process from fluid bed drying to microwave drying.

These investment decisions can be taken only if forecasts are reliable (although not necessarily precise).

The above examples concerned in-house decisions. Very often suppliers also have long lead times to order their equipment. Moulds for plastic parts are a case in point. A mould for a plastic aerosol actuator may cost $ 1 million and may take 9 months to make and bring on stream. Often, companies have run into trouble because moulds were not properly made because they were made too quickly; or insufficient time has been given to sort out problems in detail.

Major quality problems can arise. These are mostly caused by poorly fitting plastic parts or ill fitting connections existing between plastic and metal components.

Frequently it is in the interest of the pharmaceutical company to participate in the investment, or indeed to wholly own an expensive item such as a mould for glass or plastic component. In this way, the supplier has less risk and can compensate this by preferential prices and delivery conditions.

Often these issues are not realised by either the commercial or the people defining strategy. This can happen in the centre of an organisation which is far removed from the practicalities of both the supply chain and the technical fields.

It must be repeated that these issues which one might hitherto have been associated with new products can thus be shown to apply equally to well established pharmaceuticals.

Moreover, the inverse situation can also arise. Sales figures might not achieve those expected. In this case, it is possible that one might decide not to use the equipment which was designated but instead use smaller, untested equipment for economic and supply chain (cycle time reduction) reasons. This is particularly true if batch sizes have to be smaller, or more changeovers from one size box or pack to another is required. Quality problems associated with the new equipment could thus occur.

In one company a product launch was not as successful as expected. A large ampoule filling line, which had been specially bought and installed for this product, was switched to another product. Older and smaller equipment was therefore used for filling the ampoules. Unfortunately, the smaller machine did not have such a fine mechanism as the larger one. The result was that many ampoules were broken or cracked during filling. The consequences could have been dire – bacteria could have contaminated the product through the cracks, despite the fact that they were passed as sterile after making. Fortunately, the problem was discovered in-house and another means of production – in this case outsourcing, was found.

Storage space not properly foreseen for unexpected volume increases can create serious quality problems. In terms of pallet space, it is often forgotten that the space required to store components for an aerosol, a tube of cream, or a syrup, might be double or treble that of the finished pack.

Due to lack of space, one company decided to store its empty bottles outside. The bottles were properly packed and protected from vermin, dust, and dirt. Unfortunately, a freak spring condition suddenly made the temperature go from plus 14°C to minus 4°C, and then within 12 hours up to plus 8°C. Many of the bottles cracked, and although the computer showed the presence of 200,000 bottles, only 30,000 were usable – thus the supply chain was broken.

Sharing of Risks between Manufacturer and Marketing

The consequences of selling many times more than forecast, or many times less, have to be carefully stated by manufacturing and have to be known and understood by marketing.

The risks attached to these consequences have to be shared in defined proportions between the two functions. The author has seen a number of cases where marketing overestimated sales by a factor of 20 or 50. *In another case, they underestimated sales by similar numbers.*

The consequences have been very costly either in loss of sales or heavy over-investment and commitment to suppliers. As mentioned before, new and original equipment (purpose built) can take from 12 months to two years to build. New factories for raw materials take much longer.

The important point to stress here is that these lead times should be known by top management including finance. The overall commercial and financial impact of the consequences should be assessed carefully, honestly, and together. The cost of the loss of image of a company which launches a new product and cannot keep up the supply with the sales is very high. Especially this is so if the product is good and/or innovative. And as mentioned previously, clear action plans should be pre-established and known by all the people involved in the chain.

These examples endeavour to show the need and the consequences on quality, of good long term forecasting.

Agile business thinking, anticipation, and flexible capabilities can obviously help in the risk assessment and risk taking.

The identification of people involved, and the closeness of the deciders to the market, equally help in this process.

The Whiplash or Bullwhip Effect
(also called Whipsaw effect)

This is the amplification of sales or forecast variability along the supply chain.

Information gets distorted at each and every link as it travels backwards up the chain. Thus a multiplying effect takes place and stock is "snowballed" either into larger and larger stocks – or the inverse.

Patient   Pharmacy   Wholesaler   Manufacturer   Raw Materials
                                                     & Components

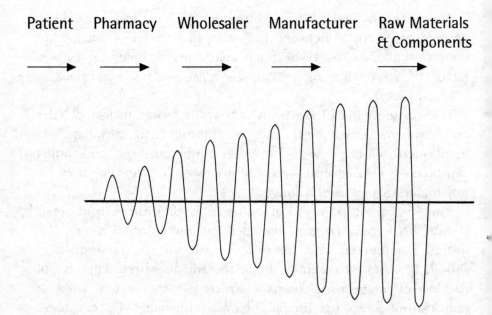

The result is that the capacity planning system is upset. Scheduling and transport arrangement become inefficient, the stock at various levels piles up.

But worst of all, customer service is degraded, resources are wasted, and revenue is lost.

The Bullwhip effect is due to:
– poor feedback from Marketing,
– misinterpretation of Marketing figures,
– poor information flow between manufacturer and supplier.
– misunderstanding of the supply chain,
– misunderstanding of stock management.

It comes from taking irrational decisions (panic decisions) and unclear, probably too rapid, and erroneous information upstream. In other words, a lack of partnership and common sharing of vision.

Medium term forecasting (Material and Resource Planning)

Somewhere in between long and short term forecasting is the vital activity often know as Materials and Resource Planning (M.R.P.) (see Figure 2.2).

This is nowadays a packaged computer programme which is either bought in or created in-house. This essentially translates sales figures into production figures and calculates the input of the various components required to produce for sales. Normally MRP forecasting operates 2-3 months in advance of actual production dates, depending on the lead times for the various components and excipients. How far it goes to scheduling the actual work place and/or the supplier, depends on the sophistication of the programme.

Its success once validated depends on:
– the accuracy of the forecast,
– the accuracy and precision and reliability of the input data,
– the reliability of the cycle times,
– the reliability of the supplier and his suppliers.

The quality input is the checking of the reliability of the data.

Short term forecasts

Short term forecasting is necessary for calculating machine loading and staff utilisation. Calculating optimal batch sizes and the number of batches to be run, without changeover, are essential features of the supply chain and its effect on quality. Any change-over implies not only loss of time but a critical quality point, where a potential mix up of product, product strength, or packs, could occur.

Thus it is in the interest of both the supply chain, quality, and of course cost, that change-overs should be minimised. However, the just-in-time principle requires small batches and frequent changes.

Therefore, the two constraints have to be balanced. This is the

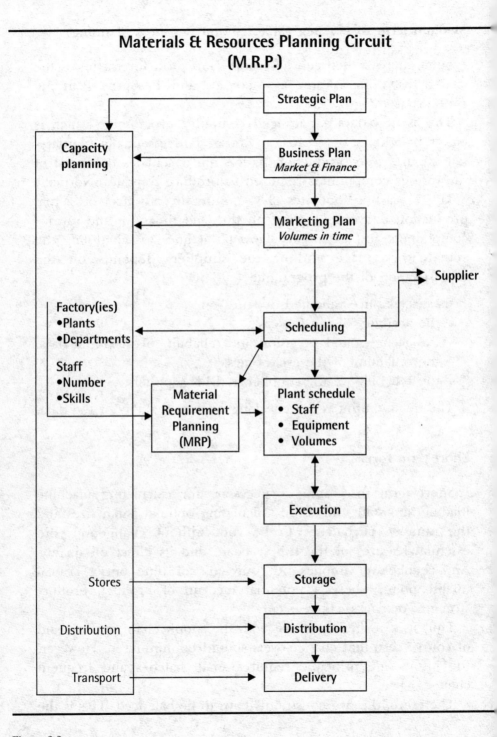

Figure 2.2

crux of successful short term planning – which is a consequence of short term forecasting.

A dramatic effect on quality due to bad short term forecasting may be illustrated by the following example:

The same manufacturing vessel was used for making a number of creams and ointments. There was a defined sequence in which the ointments and creams had to be made in order to reduce cleaning times. On one occasion the forecast for one particular ointment was changed – twice the usual quantity was required. This posed no problems, but the next batch of ointment was only finished at seven o'clock in the evening and was not immediately transferred to a holding tank.

During the night, the watchman observing the unusual fact that a batch was left in the manufacturing vessel, phoned one of the responsible technicians who told the nightwatchman to turn off the electricity which might have been left on. This was done, however the following morning the ointment had set to a hard mass, and only by overheating it could it be removed from the vessel. Needless to say the batch was lost.

Influences affecting forecasts

Although I have described the three types of forecasting and their typical characteristics, there are a number of factors which give rise to perturbations in forecasts which, if not taken into consideration, could cause either out-of-stock situations or stock mountains. In some cases the effect is not felt immediately – therefore causing problems at a later date. The following are the more important ones.

Seasonal Variations:

The pharmaceutical industry, like many other industries, is under the influence of seasonal variations. Antibiotics are used a lot more in the winter (in the northern hemisphere), especially when 'flu epidemics occur. Anti-asthmatics are sold more in the spring or whenever the pollen count rises. 'Flu vaccines are used virtually only in the autumn.

Although there is a constance in the variation, the actual degree can vary by 300 %, even from country to country on the the same continent, and also from continent to continent. Unpredictability is the key word. Although some epidemiological information may give an indicator of the size of an epidemic, it is not a reliable indicator.

If stocks are too low an out of stock situation arises unless the company is highly reactive. If stocks are too high, then one may well be left with unsaleable products or products with shorter shelf life. These are difficult to sell.

Capacity within the pharmaceutical factory is critical at these periods; not only equipment capacity, but also availability of staff. If temporary staff are used then the quality issue could become critical.

Booked capacity in suppliers is also critical because the supplier in turn might have to engage temporary, and perhaps untrained, staff.

Experience shows that when there is an influenza epidemic, medicines other than antibiotics also increase in sales. Anti-inflammatory and vitamin drugs are obvious examples, but it goes further than this because as more patients go to doctors, they are prescribed other drugs for all their other ailments.

Historical and statistical information can help to make intelligent guesses. They can calculate the quantities to be stocked, or have the necessary work in progress and which must be made at short notice. This means that it is essential to have raw materials and components available.

The uncertainty of all this costs money; either because of high stocks or because of an out-of-stock situation and subsequent loss of sales to the competitor, which can be forever. Working overtime, double or treble shifts, generally costs more. Using temporary labour might not only give quality risks and is often, but not always, slower and less efficient.

All these factors put a strain on a factory where there is already strain. Thus particular attention must be given to all quality aspects.

Somebody must take the decision as to the level of stocks to be held. Generally, it is Manufacturing who takes this decision. This is inadvisable. Instead it should be a shared decision between:

- Manufacturing,
- Marketing,
- Finance.

It is just not advisable that the factory should carry all the extra costs, thus increasing its ex-factory costs and obviously lowering its competivity.

Seasonal variation is an integral part of the pharmaceutical sales and manufacturing pattern, the supply chain integrates this. Quality-wise the variation represents challenges, where vigilance and strict adherence to Good Manufacturing Practices are essential.

New Product Launches and Pack or Presentation Variations:

I have written already about the need for accuracy and reliability. Nowhere is this more challenged than by the launch of new products, packs, or presentation variations. A single pack component if not foreseen early enough can cause delays – and often chaos. The design of a new presentation might also take longer than expected or necessary – thus having an impact on forecasts.

Although product launches are often planned a long time in advance, the actual date of launch might not be known. This is due to uncertainties of regulatory approval, pricing authorisation, or even commercial and competitor reasons. Quite often, for some countries to obtain the price from the government might take anything from three months to three years from the time the approval for sale has been obtained!

One of the biggest risks which can occur is over-optimistic forecasting.

In this case, too much active ingredient could be made and may remain too long in stock, thereby reducing shelf life. Similarly, a finished product might go out-of-date or have a very short expiry date by the time it goes on the market. It is therefore inadvisable to start manufacturing a batch or a series of batches for a product launch without first having a degree of certainty in the forecast launch date, and quantities.

Seasonal variations on new or recently introduced products, where there are no historical statistics, present an even greater challenge. The problem could be aggravated by the fact that there might not be experience of the component suppliers. They might be unique, or new to the company or product. Hence their quality has to be surveyed more closely because, for the first time, they might be asked to work under pressure.

Quotas:

Some countries have limited funds for importing goods. The licences often take a long, and above all, an uncertain time to obtain. Frequently a country obtains its import licence ten days before the cut-off date for the year's quota runs out.

Obviously, everything has to be done to make and ship the goods within the limited time. Making the product earlier would be risky because the licence might never have been obtained, or would have been granted for a different quantity. These quantities are usually absolutely precise; if 10,000 of an item are ordered, one cannot ship 10,200 or 9,800 (which might have corresponded to a batch size).

All these points show the need for reactivity and short cycle times as well as rapid information flow.

Critical Factors:

- Different long term strategic and short term forecasts,
- Accuracy and Reliability,
- Known degree of precision,
- Information Flow,
- Peaks and Trough forecasts,
- Seasonal Variations,
- Commercial Variations (campaigns)
- Pack change planning,
- New pack introduction,
- Quotas,
- Knowledge of bottle necks,
- Knowledge of Distributor's stocks.

Quality Actions:

- Team work in making forecasts,
- Common understanding of Product Supply and Commercial Constraints,
- Fast feedback system,
- Differentiation between strategic and short term forecasts,
- Updating of forecasts,
- Understanding of cost of forecasts variations,
- Rapidity and precision of communication,
- Communication throughout the chain (i.e. all suppliers),
- Action on bottle necks,
- Clear responsibility and accountability,
- Work on one set of figures and plans.

Critical Quality Indicators:

- Forecast accuracy
- Forecast deviations,
- Number of demand variations
- Size of demand variations
- Trend records,

- Supplier record,
- Distributors satisfaction survey,
- Out of stock record,
- Value of stock record,
- Stock level/demand ratio,
- Replenishment speed

3 PROCUREMENT

Procurement is the purchasing and ordering arrangements which are formalised between one company and another. It includes both price and payment negotiations.

The three current methods of procurement are:
- straight order with quantities and a delivery date or dates.
- a blanket order from which deliveries can be called as and when required or preplanned. This arrangement extends generally from 6 months to 3 years.
- the VMI system: Vendor Managed Inventory where there is a total transparency between customer and supplier concerning stock levels, actual orders, and forecasts. Here the supplier is required to guarantee availability of the goods within an agreed minimum and maximum stock level, and is accountable for it.

The choice of the system will depend on the frequency of orders and business volume.

Price. Negotiations

The price and commercial arrangements are the linchpins of good supply chain management. With each and every supplier in the chain, it is vital to agree in terms of:
- Price,
- Payment,
- Revision clauses,
- Delivery conditions,
- Lead times,
- Order quantities,
- Reaction times,

 – Total capacity,*
 – Maximum capacity. *

Constraints, or undue pressure, on any of these factors can lead to major quality problems and/or supply chain breakdown.

The most common problem is price. If the pharmaceutical company exercises too much pressure to force the supplier to lower the price – then this can drive the supplier to the wall in which case he will not, or cannot, deliver to the required quality and to the right specifications.

Often this can be seen where, for instance, the quality of the paper for cartons has been reduced without the supplier informing the user. Another example is the quality of the plastic used for a bottle cap which was changed, thus making it weaker. It is true that problems of this type could be avoided by tighter specifications ; however there are a multitude of examples where this can, and does, occur.

Another example where analytical control has little means to quantify quality is in flavours. If a manufacturer switches the flavour supplier because the new source is cheaper, it could find itself having problems. Sophisticated analytical means do exist to differentiate betwen artificial and natural aromas, but not necessarily between one natural and another natural source.

A manufacturer must agree "honest" prices which the supplier can guarantee to maintain. He should, at the same time, be able to show that he makes enough profit to keep his business in good order.

This transparency, where mutual benefit is apparent, is necessary to create a reliable supply chain.

Revision clauses in the commercial contract should allow open review of prices. Also it should allow competitor quotations and benchmarking to obtain *mutual* benefit. This lays the foundation for Partnership.

Payment conditions, and indeed the contract between supplier and the manufacturer, must be such that he cannot turn round

---

* although capacities are considered as more technical than commercial – they are still a vital part of the commercial contract.

saying "I supplied late because you paid me late" or "you paid me late, so I had to use a cheaper raw material".

On the other hand, a supplier cannot be expected to specify, name, or guarantee all his sub-suppliers. It is his responsibility that they should deliver to his specifications.

Supplier lead times

Knowing a supplier's lead time is essential to the supply chain, it being more important than his delivery time. In fact, the two are often confused and thus create problems. The lead time is the time that a supplier takes to deliver goods from the time he receives his order, i.e. it is part of his own supply chain.

It includes such factors as:
– Order processing time,
– Stock,
– Purchasing item lead times,
– His manufacturing cycle time,
– Delivery time.

It is necessary to understand this very well in order to avoid the surges in production that lead to quality problems.

Very often an item is ordered and is delivered within 10 days. This leads one to assume that the lead time is also 10 days. Yet, if two or three orders are placed in a row, or a much larger order is placed, one finds that the delivery time is 45 days!

The problem is that the supplier delivered from stock up to a certain order level and at a certain frequence. Beyond that, he has to re-order from his own supplier thus causing the consequent delay. Sometimes, the difference is much bigger than the above example and then really big out-of-stock problems can ensue.

The example above also shows the reactivity problem; one is accustomed to getting fast deliveries. As a result one is led to believe that the supplier is reactive – yet, when the crunch comes, his delivery time lets you down.

Anticipating the unexpected, which is agile business thinking, can often avoid this.

Reactivity

To be really reactive, a supplier has to be able to deliver fast, even when he does not have the item in stock. His own cycle time and his own supplier arrangement must be such that he can radically and rapidly change his supply time.

Plastic parts which have to be moulded in large series in order to reduce the changing of the moulds in the presses, are a typical item where there are long lead times. For similar reasons, another example is glass bottles.

The pharmaceutical manufacturer who really wishes to have flexibility and reactivity for this sort of item, is advised to have two proven suppliers. Or he had better get his forecasting right.

With printed items, such as leaflets or cartons, such difficulties should pose less of a problem. However, if the price arrangements are too tight, a supplier will not easily switch in a hurry from a large production run to a sudden smaller one. His margins would be quickly reduced. Quality problems can also occur if one wishes to switch very rapidly to new suppliers of printed items. In the pharmaceutical industry, artwork requires both careful checking, and double checking. A supplier either not used to the industry, or not used to producing the specific items, might find it difficult to deliver the quality product on time.

Critical Factors:

- Price agreements,
- Commercial conditions,
- Components, see page 10
- Transparency of interactive database,
- Communication (where no interactive database exists),
- Understanding financial constraints,
- Common agreement on lead times,
- Storage conditions and constraints,
- Defined transport arrangements,
- Defined delivery arrangements,

Quality Actions:

- See section on Supplier Relationship.

Critical Quality Indicators:

- See section on Supplier Relationship.
- Time and quantity delivery records,
- Price variations.

4 | *SUPPLIER-CUSTOMER RELATIONSHIPS (PARTNERSHIP)*

This chapter is principally concerned with the relationship between a supplier and its pharmaceutical manufacturer – the traditional supplier-customer relationship. However, in line with the desire to look at the pharmaceutical product supply chain in its entirety, the relationship between the pharmaceutical manufacturer and *his* customers and partners is also examined. The concept of ASSCENCE as a management principle, system and frame of mind is described, which applies both to Suppliers and Customers.

ASSCENCE

Modern customer or partner* relationships should be based on a behaviour pattern or system which this writer has named ASSCENCE. It is an agile management technique. It is the way a supplier can contribute most effectively to the customer.

It stands for:
ANTICIPATING, **S**USCITATING, and **S**ATISFYING **C**USTOMER **E**XPECTATIONS and **N**EEDS in a **C**OST **E**FFECTIVE MANNER.

It is rather a heavy definition, yet it endeavours to be complete. It differs from Efficient Customer Response (E.C.R.) by being far more limited and focused in scope, but in essence is part of the same philosophy.

The word "Suscitate" is old and rather out of use (it should be resuscitated!). Its root is the latin citare, meaning to excite, to stir up, to raise out of inactivity, to quicken, vivify, animate. In

---

* It may be useful to remind ourselves of just some of the partners involved in the supply chain: Regulatory, Marketing, Medical, Maintenance, Customs, Transporters, Quality Control, Warehouse, Distributors, Finance, etc...

management language, it means to stimulate, provoke thought, and motivate.

A supplier should, with the customer, anticipate the needs or the likely needs of the customer. The importance here is the anticipation made in partnership with the customer. This obliges sharing of thinking, sharing of responsibilities, and sharing of objectives of both (or more) parties. It is an integral part of agile supply chain management.

Figure 4.1 captures the concept

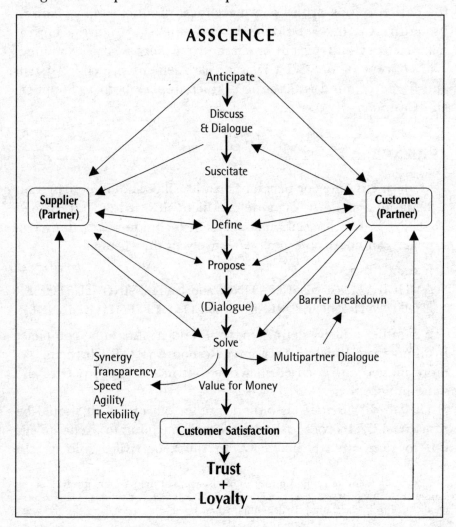

Figure 4.1

The system assures synergy in rapidly finding ideas and solutions to problems. Both partners should be committed in creativity, innovation if need be, and curiosity in finding the best and most cost effective solution.

The supplier who blindly supplies whatever the customer wants, or thinks he wants, is not doing his job fully. His job is to find the most cost effective answer to his client's needs, and/or propose alternatives.

Enormous savings can be achieved by defining requirements better and aligning capabilities with needs. Indeed, often the success of Outsourcing is due to the supplier finding better and cheaper ways of achieving the customer's goals. Working out together optimal timing and scheduling is also necessary to obtain the best commercial solutions.

It avoids the pitfall of believing that only the customer knows best what he wants. Very often, the supplier can think of a better solution.

This is all common sense, but not often practiced either out of laziness, fear*, habit, timidity, or false economic reasons. The final decision point of course does not alter, and is in the hands of the payer.

Obviously this way of working takes up some time and effort, but it is quickly repaid.

ASSCENCE is Total Customer Service i.e. giving full satisfaction to the customer with the best value for him.

In agile manufacturing where we talk, amongst other things, about mass customisation and the way to achieve it, that is by flexibility,reactivity, and with speed, ASSCENCE is a key and vital element. Here the anticipatory role comes in play, as well as finding rapid solutions.

Very often a supplier has more than a single customer or decider for his product or services. In this case, the system

---

* With new relationships and/or sensitive subjects which could require a high level of confidentiality, precautions must be taken in as far as divulging information. Legal contracts may be required to safeguard against loss of intellectual property.

New relationships also require progressive openness while trust and confidence is being established.

naturally involves all the players. Barriers which so often hinder good decisions, choices, contracts, prices, etc... are helped, by this system, to be broken down.

With this method of working a supplier or partner can never turn round and say "the customer never told me" or "the customer did not know what he wanted". It is totally unacceptable in a win-win partnership to blame the other. Responsibility must be shared in any supply situation.

ASSCENCE is a very powerful facilitating method, in establishing Trust (see Chapter 14) which is a necessary element in any long term partnership.

ASSCENCE is really a frame of mind and an ingrained way of working, showing a positive, constructive, transparent, and willing attitude between two (or more) partners, each of whom is seeking mutual and long term benefits. It is an important management principle, applicable in most relationships. It is nothing new, but it endeavours to bring together and highlight a way of working which is of utter importance in the application of modern management techniques.

Partnering and Partners rather than Suppliers

The notion of Suppliers implies some subservient role vis a vis the manufacturer or purchaser. As it is mentioned throughout this book, the relationships should be transparent, honest, and ethical, with a common goal between the two or more companies involved.

A supplier has generally competencies which complement those of the manufacturer. He generally knows his particular part of the total process better.

Therefore the word "partner" rather than supplier should be retained as it confers a degree of equality between the two (or more) involved.

The word Supplier should be used more in logistic terms, for after all, everybody is a supplier of something to someone else.

The relationship between the manufacturer and his supplier :

There are two basic types of relationships between the pharmaceutical manufacturer and his supplier: formal and informal. Very often, both modes will operate. The formal relationship consists of purchase orders, purchase forecasts, delivery, and other commercial agreements.

These formal methods are more or less elaborate and detailed according to companies', suppliers', and native practices. Whereas we might see a simple one page order in the U.K. or France, for a particular item, its cost, payment and delivery terms ; the same information can take seven pages between two American firms or between an American and a European supplier.

These facts are very relevant to quality aspects of the supply chain. The level of precision stated and understood can create serious problems.

Some things the manufacturer might take for granted, feeling that they do not need to be specified or clearly stated. However, the supplier might not understand, or in some cases might not want to understand.

We can take the example of the siliconing of glass vials used for antibiotics. The pharmaceutical manufacturer just stated on the order "siliconed vials". When the vials were delivered, it was found that the siliconing was not done in the way the manufacturer expected. Perhaps the layer of silicon was not sufficiently thick or there was variability between one vial and another. Because the manufacturer was unaware of this problem, he did not think it necessary to specify the grade, thickness, and allowed variability.

Another example concerns antibiotic vial rubber stoppers. For years, a certain grade of rubber was supplied to a factory and although the composition of the rubber was defined, its making and processing were not.

One day, it was noticed that the seal between the vial and the rubber was not as firm as before, and could therefore allow ingress of bacteria. This was potentially a very serious problem. In fact, the rubber on autoclaving became harder and less supple. It turned out that the manufacturer of the plugs had changed his process, without informing the factory who were using the plugs. The lack of specifications relating to the manufacturing process was the cause of this quality problem.

Informal and long term relationships

Informal and long term relationships, and the mutual understanding between manufacturer and supplier, is one of the keys to avoiding quality problems. Detailed visits to each other's manufacturing establishment, knowing the people, knowing the processes, knowing the analytical procedures used, may well be heavy investment in resources, but are essential to obtain the required quality of goods. This is the basis of Partnering.

No amount of written contracts and agreements can substitute for this kind of relationship, which is a form of partnership. From this, trust is progressively established and thus risks are reduced. (see in Chapter 14 on Trust).

Speed, agility, and reactivity are far more easy to obtain.

When it comes to equipment suppliers, a similar relationship should exist in order to get rapid after sale service in case of a machine breakdown.

The question to ask is that in case of an emergency, after how many hours does a machine constructor arrive. Hopefully, it is as soon as possible. However, the speed with which suppliers react, reflects both the quality of the relationship and the quality and efficiency of the supplier.

Supplier as Supply Chain Adviser (Consultant)

It is rarely realised that suppliers both of equipment and packaging items and excipients are in a potentially privileged position to further the supply chain of their customers i.e. the manufacturer.

As they have generally a number of clients, they can learn from their efficiencies (and inefficiencies), especially from other than pharmaceutical industry in reducing cycle times, avoiding waste, and using optimal staff both for the manufacturing process and for the packing lines.

Furthermore if they supply other industries, which might be much more efficient than the pharmaceutical, then they can transpose what they have learnt, without endangering the competitivity of the more efficient firm. They can provide unexpected needs and unfulfilled possibilities to their clients. This is what Stephen Blackledge calls the Competence Web.

Let us take the example of a manufacturer of mixers for tablet granules. One of his objectives in selling his machines is to persuade the buyer that his machine induces only low losses and has a short cycle time. He might argue that the granulating plus cleaning time takes two hours and 30 minutes whereas his competitors' machine takes 3 hours and the manufacturer's cycle time for a product may be 15 days!

The point is not the relative difference between the two machines but their insignificance compared with the whole cycle, and that the machine maker knows the length the cycle and therefore, if he wishes and has the ability, he could very usefully advise the manufacturer on how to reduce his overall cycle. This is a very important sales advantage which very few machine manufacturers make use of. By showing the economies in reducing the cycle, he could persuade his potential customer more easily.

Two problems must be considered in this scenario, one is that of confidentiality and the other that of supply chain planning and scheduling competence.

But by careful observation, learning, and the diplomatic use of the knowledge gained, (especially from other industries) – suppliers could make themselves much more useful.

A similar situation could occur with a carton or glass bottle manufacturer. By observing short and efficient cycle times in some firms, the advice that they could give to others, based on this knowledge, could be a very attractive sales argument.

Suppliers too should consider their trade in holistic terms – a machine maker is not only an engineer but is selling an efficient machine for an efficient production cycle. A bottle maker is contributing similarly in improving his partners competitivity! (see paragraph on the Stonemason on page 225).

Sharing of competencies among companies, not necessarily competitors, can give tremendous advantages to companies, at very little cost. These competencies are not only in techniques of supply chain management, outsourcing, engineering, work study, and improvements matters, but can extend to general management matters such as training, costing, benchmarking, and quality assurance.

Relative Size of Supplier

Some of the most critical quality problems concerning suppliers revolve around the size of the suppliers, and notably the relative size between the manufacturer and the supplier. Similar problems can occur with newly established or young firms.

If excessive leverage is exercised by a manufacturer on his much smaller supplier, he could drive him to bankruptcy and thus would have to find another supplier. Alternatively he might drive him to produce lower quality goods. It is important that a win-win attitude is built up by the more powerful organisation.

A situation in reverse might occur when the pharmaceutical manufacturer requires modest quantities of goods from a very large supplier. This supplier, in turn might find the business unattractive and therefore not pay sufficient attention to the purchaser.

The pharmaceutical manufacturer often requires small quantities with very high specifications. An example of this is rubber components for aerosol valves. The requirement of a large international group may be half a ton of rubber per year. This is equivalent to half a dozen lorry tyres. This order might be trivial to a large reputable rubber company such as Dunlop.

Equally, a pharmaceutical company might express its requirements for millions of tiny washers or whatever, each having very tight specifications, each having to perform critical quality roles. Although the price paid could be hundreds of times higher than the tyre – the problem still stands. It requires skillful negotiation to allow a win-win arrangement under these conditions.

Similar cases can arise with small plastic and metal components. The case of sodium citrate is being discussed on page 98.

Young firms:

As mentioned previously, critical quality problems are more likely to surface with new, start up, or young firms. But, despite this, it is very much in the interest of manufacturers to help them to develop and to develop well. Such firms are usually much more open to advice. They are conscious of quality matters and are generally eager to learn. The investment by the manufacturer may be quite heavy in time, advice, counselling, and even sometimes in advancing funds. But the return can also be very satisfactory for both – provided long term arrangements are set up.

There are risks of course, as with all new ventures. For instance, it is very important that the manufacturer should not be, or become, the unique client of the supplier. This can very easily happen because it is comfortable for the small supplier not to have to go out and find new clients – thus allowing him to concentrate his efforts on his main client.

The problem arises if the manufacturer has financial problems, or finds a better or cheaper alternative supplier. In this case, the small young firm could become bankrupt within months if not weeks. An insurance against this is that in any contract it should be specified that the manufacturer should not represent more than 65 % of the supplier's business. This might be against free trade laws in some countries. Yet in other countries, the opposite system might operate. In this case, a supplier must have more than one customer, otherwise a monopoly law is infringed.

One must be careful with dogmatic purchasing policies and avid cost cutting buyers, who feel they are judged by the yearly economies they make for their company !

Whilst it is becoming more and more important to reduce purchasing costs in order to remain, or become, more competitive, each and every supplier must be treated according to individual situations. Scarcity of goods, scarcity of other good suppliers, size of firm, management attitude and structure of the supplier, are just some of the considerations which could or should influence the policy vis a vis a supplier.

Across the board sweeping instructions, such as "we need a 5 % reduction in cost every year" can lead to disastrous results either in terms of quality, reliability or both.

The relationship between the manufacturer and his customer:

In this case, the supplier is the pharmaceutical manufacturer and the client is the wholesaler, hospital, clinic, doctor, or even in some countries, the patient. The quality aspects of this relationship are discussed throughout this book. For a manufacturer to be a good supplier and partner, for him to be trusted by his client, if one single aspect of the relationship is to be mentioned, it is – reliability.

It is obvious that the manufacturer must know what and when the customer wants something, and what quantities he requires. Here the information flow is vital. Informing a customer of delays, late deliveries and quantity changes, are essential to maintain trust and keep loyalty.

The concept that the manufacturing system INCLUDES or INVOLVES the customer is becoming more and more accepted. The level of interface between manufacture and customer has also to be defined – sometimes, the workshop or manufacturing unit should be in direct contact with the customer.

The staff of a manufacturing company can be far more easily motivated to exceptional efforts, if it can identify with the customers. The physical proximity of staff to customers helps a great deal in being involved with the customer. Many an out of stock situation has occurred because of staff, and where management were not embracing and identifying with customer needs.

Client Included Manufacturing Systems (CIMS)
Towards Agile Manufacturing

Customer Value

CUSTOMER

Systems	Process	Measures	Objectives
QA	Conversion	Activity base	Cost
Teaching	NPI	Costing	Speed
Admin	Engineering	Accurate	Quality
A/C	Procurement	Focused	Anticipatory
I.T.	Hum. Rel.	Performance	Service
		Time Value	Reliability

PRODUCT FLOW

CUSTOMER

This concept incorporates many of those in ASSCENCE (page 49).

The figure 4.2 on page 59 shows the Client Included Management System (CIMS) which is used by some forward looking organisations.

The relationship between the manufacturer and its sister company

This is a particularly important aspect of the supply chain in all large pharmaceutical companies whether national or international.

It is important for the following reasons:
- it might represent a large or major part of a manufacturer's client base,
- it is "controllable" in the sense that it is within a single corporation's ambit,
- it is often complex due to different countries' and different markets' commercial and regulatory requirements,
- it is sensitive because a sister company might be more demanding than an outside client, and there could be internal company politics.
- objectives and priorities of the manufacturer might not be totally in line with the sister company,
- multiple market forecasting might completely upset production planning.

It is usual, for these reasons, to set up internal supply chain performance measures (see page 215), which will become part of the judgement criteria for rationalisation and sourcing decisions. The quality issues revolve around different registration parameters from one country to another (see chapter 12 on Regulatory Issues). However, the most important and difficult issues concern product, pack, and quantity changes.

The technical aspects of these issues are mentioned in New Product Launches and Pack Changes, (see page 65). However the supplier relationship aspects are similar to those with exterior suppliers. Close contact, clear understanding of requirements, understanding of product constraints on the one hand and marketing constraints on the other, are key elements to avoid conflict. Personal contact, knowing who to contact rapidly, and knowing who are the decision takers, are essential features to insure the smoothness and agility of the supply chain.

These relationships should exist not only between the export or planning departments of the two companies, but should also exist between Quality Assurance managers and transport managers.

The more direct contacts there are, and the fewer intermediate people that get involved, the faster and clearer the messages get across. Hierarchic and administrative methods of working both slow the process down and also impede the supply chain.

Often regulatory people and sometimes staff from the medical department have to be involved in changes or specifications. In order that all should simultaneously be aware of the situation, regular meetings between manufacturer and the sister company client is recommended, even if it means crossing the atlantic (provided of course that the turnover is, or is expected to be, large enough).

Many wasteful hours of telephone discussions, e-mails, and faxes can thus be avoided.

See Client Included Manufacturing Systems in previous section, and agile management techniques.

The following list of critical factors, quality actions and critical quality indicators are mainly directed at the supplier relations. There are a number of items which can be equally important in the case of customers.

Critical factors:

- Starts with good forecasting (see section on this),
- Rapid and decisive information flow re change of:
    - requirements (specifications and modifications),
    - quantities.
- Sharing* of both responsibilities and of risks,
- Knowing his capacity,
- Knowing his capability,
- Technical know-how,
- Technological complexity of supplied item,
- Clear contracts,
- Mutual trust.
- Sharing vision,
- Partnership.
- Win-win attitude.
- Audits, supplier training – 2 ways, their people in factory and vice versa
- Competitivity.
- Know people by person – not just telephone.
- Know mutual constraints.
- Information technology capability.
- Flexibility.
- Speed of reaction.
- Benchmarking with other suppliers.
- Critical size of supplier.
- Financial position
- Direct computer and/or phone line – for precision and speed.
- Knowing their supply problems and the risks attached to these.

* This is generally one of the weakest points in the relationship between pharmaceutical companies and their suppliers. Often at the Development stage "scientists" commit themselves to one supplier without prior proper business negotiations. This is usually done to gain speed. However, in the long term, prices are difficult to renegotiate and the companies negotiating position is weakened. On the other hand, the supplier has no risks to take and is hence in a comfortable position "forever".

– Social climate.
– Transparency – avoidance of misinterpretations.

Quality Actions:

– Clear agreed specifications
– Audits, not only Quality, but also Management and all other critical factors,
– Shared Quality Control Methods and levels of acceptance.
– Train suppliers staff.
– Train suppliers Quality Assurance and Quality Control staff.
– Keep suppliers to a safe minimum.
– as in Forecasting, see page 41.

Critical Quality Indicators:

– Forecast deviation monitoring,
– Reliability of their Equipment/Capacity, knowing their bottle-necks
– Their Quality Control systems and rigour.
– Knowing their financial condition and its evolution.
– Stock levels internal, and held by supplier.
– Delivery Records.
– Out of specifications deliveries.
– Type of specifications errors.
– Speed of reactivity to errors.
– Number of suppliers for same item.

5 | MARKETING AND NEW PRODUCT LAUNCHES AND PACK CHANGES

The Marketing division or department in a pharmaceutical firm is the key element of the supply chain. It is the commercial link between the manufacturer and the client or patient. Yet it is often forgotten, or taken for granted in the chain. Manufacturing frequently assumes that marketing is a reluctant partner of the link.

Marketing is the initiator, often the driver of the chain and it must be considered as a very close partner in order to get the full benefit of the economics of the chain. Not only does it define specifications, at least in broad terms, but is also is responsible for the Forecast, their modifications, and also a whole set of other specific market or commercial requirements. It is, of course, generally responsible for selling the goods.

In the pharmaceutical industry the specifications of the actual products are defined by Research & Development. However, there are a number of other specifications which in many cases, are guided or under the control of the regulatory authorities, but are decided by Marketing.

Examples of these are:
- text layout,
- sizes and shapes of packs,
- colours,
- leaflet layout (text, size, shape),
- shape of bottles,
- tablet strip sizes and shapes,
- bottle caps (shape, embossed, language),
- size of sample pack,
- label, size, transparent,
- etc,...

Now, the important thing to realise is that in a partnership situation, some of these specifications are negotiable. And it can often be easily demonstrated that by slightly altering, for instance, the shape of a box, one can achieve considerable economic benefits as well as speed in supply. This is obviously advantageous both at the factory level and the carton supplier level and thus affects the supply chain and its efficiency.

The importance of closer relations between Manufacture and Marketing is illustrated by the following example.

A company wanted to harmonize its tablet design in order to reduce or avoid changeover times – a changeover of punches and dies can take 3 hours and this would be a necessity regardless of whether the run is for 3 hours or 3 days.

It negotiated with the technical people of the sister company who objected, saying that their marketing department would not like it. In fact, they did not consult their marketing department and it was the supplying company who finally got in touch with Marketing and obtained their agreement, without the slightest problem.

It is evident that in a multimarket and globalised manufacturing situation, standardization gives both flexibility and economic benefits. As margins in the industry become narrower, marketing people will become more receptive to arguments concerning pack modifications and other mutual benefits – which give overall benefits to a firm. In agile business management, it is vital to understand the overall cost of a product, a functional or partial view of costs or benefits is economically wrong.

Packs and packing specifications, their harmonization and standardization impacts on quality by simplifying:
- changeovers from one pack to another,
- easier identification,
- uniform tablet sizes,
- carton weight and size,
- colours.

Critical factors:

- Mutual understanding of requirements,
- Precise understanding of requirements,
- Good communication and simple explanations,
- Pack specifications,
- Order quantities,
- Pack quantity groupings,
- Pack differentiation or colour differentiation for different dosages,
- Understanding of costs, in an overall and holistic manner,
- Understanding of margins,
- Rapidity in decisions,
- Special pack requirements,
- Alignment of Objectives,
- Goods returned from Market procedures,
- Frequency of pack changes.

Quality Actions:

- Schedules planning reviews,
- Standardization,
- Simplification of packs,
- Robustness of packs,
- Customer feed back process,
- Pack /product improvement programme,
- Agreed and approved transport arrangements,
- Agreed and approved distribution arrangements.

Critical Quality Indicators:

- Cost assessments, (of different packs of the same product)
- Ex-factory costs,
- Customer satisfaction indices,
- Complaints, – justified and – non justified,
- Medical representatives reports and comments,
- Transport breakage or damage records.

– Goods returned: – quantities,
                           – reasons,
– Number of pack changes.

Marketing and Packs:

Manufacturing is there to provide what marketing wants! This was the logic in the past and still is to a very large, but not total extent, today.

Marketing very often states its pack requirements based on copies of something that already exists among its competitors. Rarely does it invent, and rarely is it creative.

Manufacturing, similarly, is generally not incited to be creative and propose something new or different. (There are obviously exceptions to these two statements – which only prove its truth. Creativity however, is one of the most under-used quality which employees have; the climate for their expression does not exist in most firms).

As an example, let us take the case of a country where there is a practice that an injectable antibiotic powder is sold with an ampoule of distilled water which is to be used as a solvent.

Marketing wanted a long narrow ampoule, insisting that this was what doctors and nurses wanted. The vial next to it in the box was 2 cm shorter. The two items together in the same box either made the vial shake in the box or the box required a carton insert to avoid the vial floating in the box. When it was shown to Marketing that their margins could go up by ex-factory costs going down, they agreed to make a trial by a few hospital nurses to see how they liked using the shorter ampoules. The nurses preferred the shorter ampoules and the pack was quickly changed. The quality of the pack was improved because slimmer ampoules broke more easily during transport.

The size of foil or blister strips is often a point for discussion. Marketing want either a large surface where they can print a lot of text; or they want a certain shape, which production does not have. Often, these requirements need a lot more aluminium which is costly. They can also reduce production capacity by as much as 50 %. Having carefully explained the cost implications (perhaps by avoiding having to spend half a million dollars for a new machine) – Marketing soon changed its mind. With modern concern for the environment, where packaging materials have to be retrieved after use, the quality advantages soon become apparent too.

We must not forget that very often the supply chain ends with the return of used packaging items!

Robustness of packs, quality and thickness of cartons, impinge on the quality of the product.

The supply chain is adversely affected if a batch of syrup is returned because the corners of the cartons get crushed. Reordering of new cartons plus repacking which can take 4 weeks, could create a 6 week out-of-stock position!

Caps on bottles and on tubes is another subject where user input through marketing improves both quality and the supply chain.

For efficient high speed machines, the cartons have to be glued, which furthermore gives optimal inviolability. This, however makes it more difficult for the customer to open. So a compromise between ease of use, patient safety, and the cost of the carton has to be reached to obtain the required quality. This compromise should be agreed between Marketing, the carton supplier, the machine manufacturer and the factory.

If supply chain and quality problems are to be reduced, all these partners should be involved early in the package design.

Proper and optimised grouping of cartons in outers is again necessary for quality reasons as well as for economising on storage space and for material handling reasons.

There was once a factory which decided to have one standard outer carton for most of its products. Marketing wanted different quantities grouped for different products. The result was that most of the cartons were filled with wadding, which added to the cost of packaging, and giving a most unsatisfactory image of the company to the wholesaler. Some of the products got broken in transport – quality again. The factory very soon found the use of different cartons for different products much more economical. Transporting half empty cartons filled with wadding across half the world is not good supply chain practice!

All these examples show the importance of Marketing and Manufacturing getting together as soon as possible to work out the optimal solution for economic and quality reasons. As mentioned earlier, some supplier involvement early is equally desirable.

It is perfectly feasable to argue that one can reduce costs if packs and/or order sizes can be standardised and optimised. Marketing will listen to these financial arguments, realising that margins can be increased and order times reduced.

New Product Launches and Pack Changes:

New product introductions and pack changes are being discussed in virtually every chapter of this book. However product launches, or relaunching and new packs, are such a current business that it requires a section by itself. Since the interface with Marketing is probably the most important factor in the launch of a new product or a pack change, it seems appropriate to include it in this chapter.

It is a vital competitive feature of a firm, to launch products rapidly, smoothly and without interruption of supply – after launch. Agile manufacturing should assure the capability to respond to unexpected demand by having adequate stock and/or capacity.

It is said elsewhere that new packs should not necessarily be considered as part of the supply chain because of the exceptional nature of the event. But as companies rationalise and concentrate their production activities, factories have to deal with more and more products, and more and more markets. Consequently, in large companies, new products and new packs come out as often as once a week.

Therefore a supply chain framework has to exist, which should be similar to the usual supply chain. However it can have special features which could also present potential quality issues which might impact heavily on business.

The key features in a new product launch are as follows:

· Forecasts,
· Definition of requirements in term of   packs
                                             sizes
                                           specifications
· Spare capacity: – in equipment
                        – in space
· Lead times for follow up quantities
· Market feed back
· Reliability of Sourcing
·   »   of Process
·   »   of Equipment
· Trained Staff

As one can see, the top of the list is forecasts, and as already discussed in the chapter on forecasts, the problem is obtaining reliable forecasts for new product launches. Whilst Marketing might be unable to define accurately launch dates and forecasts (because there are many factors outside their control), they remain the best source of information, and thus communication between manufacturing and Marketing is paramount.

One of the key success factors between the manufacturing and Marketing interface is to give *Realistic, Reliable,* and *Acceptable Delivery Dates* – from the date of final agreement on price or whatever the last hurdle is.

Generally this date has to be *Negotiated* for *Mutual Compatibility* and *Satisfaction.*

Although quantities are usually agreed in advance, Marketing might want to change these quantities during the period of time in which one awaits the final decision. Unless these quantities are clearly communicated to Production they might not be available at the required time. This is especially true if capacity is already reserved for existing customers.

Samples used in large quantities prior to, and during a launch, have similarly to be forecast and manufactured in time. If a launch is delayed, a market may choose either to continue or reduce sampling. This information has to be constantly reviewed between Marketing and Production.

The example given on page 209 concerning the length of tubes for creams can be considered in the context of a new product launch. Marketing has to approve the new pack and also prepare the client for the change, otherwise the client is not satisfied.

Sampling or samples are often considered by Marketing as something simple and easy. But it must be planned, procured and manufactured, in exactly the same manner as a sales pack.

As the sample could be slightly different from the sales pack, it might have to be made on smaller, less automated, and hence slower equipment – therefore its lead time might be much longer.

New Product Introduction and Regulatory Requirements:

Regulatory departments in some companies are in close communication with both logistic and commercial departments in order to speed up the process of launching a product after having obtained approval. Therefore it is important to consider the regulatory department as part of the early supply chain team.

Pack Changes and Regulatory Groups – Agile Regulatory groups:

Specialized Regulatory Groups that serve manufacturing can be set up – because of the changing nature of pharmaceutical manufacturing to a commodity orientation, there is an increase in activity in post-approval regulatory affairs. Consequently, some companies are developing regulatory groups who specialize in serving the manufacturing sites as opposed to research and development.

The organisation of the Regulatory department is vital in all pharmaceutical firms. Their integration between Logistics, Manufacture and Research and Development is a strategic issue, both for fast launches and subsequent smooth supply chain management. These are typical examples of agile manufacturing techniques.

Critical factors:

- Forecasting,
- Forecasting update,
- Date of launch,
- Specification definition,
- Lead time,
- Production cycle,
- Artwork approval,
- Sample pack,
- Reliability of suppliers,
- Reliability of equipment,
- Reliability of process.

Quality Actions:

- Clear Specification,
- Pre market testing,
- Drawing conclusions from Marketing intelligence,
- New Supplier Audit,
- Process Validation on final product,
- Equipment   »   »   »   ».

Critical Quality Indicators:

- Time for Artwork Creation,
- »   »   »   Approval,
- Launch date accuracy,
- Time given for launch.

6 | NEW PRODUCT DEVELOPMENT AND DESIGN PARAMETERS

New products are the lifeline of most companies, this is especially true in the ethical pharmaceutical industry, where new molecules presentations and forms are brought out every day – unfortunately diseases are far from being eliminated.

Agility, speed in the design and development of products are vital competitive techniques, in a world where changes in prescribing habits and patient acceptance criteria are occurring rapidly and often unexpectedly.

Development and Marketing Information Flow

It has been stressed that the market "pulls" the products from Research & Development whilst previously Research & Development & Manufacturing pushed what could be developed and made.

This obliges Research & Development to let Marketing know what was being developed and why. Therefore the information was also one of "push". Production similarly pushed what they were capable of doing in terms of presentations, devices, dosage forms, combination packs, line extensions, etc...

Whilst Marketing and pure Research i.e. discovery of new molecules, keep one another closely informed, this is not necessarily and universally so in the development of technical fields.

In the present situation, Marketing must be constantly made aware of what developments can be made, what capabilities manufacturing and production has, what new technical and developmental work is done by the competition and other development organisations. Otherwise Marketing will be aware only of the competition once the product is on the market or, at best, at late clinical development stage. Looking outside the industry for innovative ideas, is also to be recommended. The pharmaceutical industry is far too inward looking and conservative. Many other industries are far more client oriented and user conscious, and many lessons from them could be learned.

This is an issue not fully realised by all companies, especially where resources are limited or constrained. Yet many ideas could exist which Marketing should very much welcome. A clear information system must be in place to assess regularly and fairly, the new ideas emanating outside the Marketing department.

This is similar in concept to the Supplier working out with the Customer what is the most cost effective solution to a requirement. (see ASSENCE page 49).

Early supply chain involvement in aligning needs and capabilities can greatly influence product development in assessing better the cost and the quality of the actual final new product.

Supply Chain and Design

The supply chain is established at the Design stage of a new product, whether it be a new chemical entity (NCE), or a line extension (or indeed a new dosage pack). Ingredients, actives or excipients, will have to be sourced, purchased, stored, quarantined, and quality controlled, before the product is put on the market, as stability batches and clinical supplies have to be made. Packing materials and medical devices follow a similar pattern.

Yet the supply chain is very often forgotten, or at least neglected by formulators, who obviously tend to concentrate their work on the stability, efficiency, elegance, and other technical aspects of the product. Agility is not something that they have, as yet, learned. The product is firstly made in small quantities, maybe half a kilo or up to 2 kilos. Subsequently, pilot scale batches are made of quantities ranging from 5 to 20 kilos (see later).

The product then has to be "industrialised" i.e. it has to be made in larger quantities efficiently and in a cost effective manner. Sometimes specific equipment will have to be used either for the making or the packaging of the product, or both. It is all too easy to forget that many of the decisions taken during the very early stages of product development have a profound effect on the product's supply chain. This is simply by the very nature of the pharmaceutical industry which has long development leadtimes and is highly regulated

Figure 6.1, shows the time interrelationships between the supply chain and the development cycle. It also illustrates how closely linked the two are. It is also important to pick out the registration stage of the product development cycle because it is at this point that all the decisions taken regarding the final physical characteristics of the product and its primary pack presentation are recorded by the regulating authorities. That is not to say that decisions cannot be changed after this milestone has passed. However, it should be recognised that any variations will require extra resources for additional regulatory filings, thus adding to the development timeline.

This chapter is therefore concerned primarily with aspects of product design before the registration cut-off point. The other decisions which revolve around the final market presentation pack, and the physical introduction of the new product into the market place, are covered elsewhere in this book.

With the importance of the registration cut off point firmly in our minds it is worth looking in greater detail at several of the key decisions critical to the supply chain that are taken during a product's design. Design is the foundation of any product – if it is wrong, then the product is wrong! These decisions relate to the *design parameters* of the pharmaceutical product and include the following subjects:

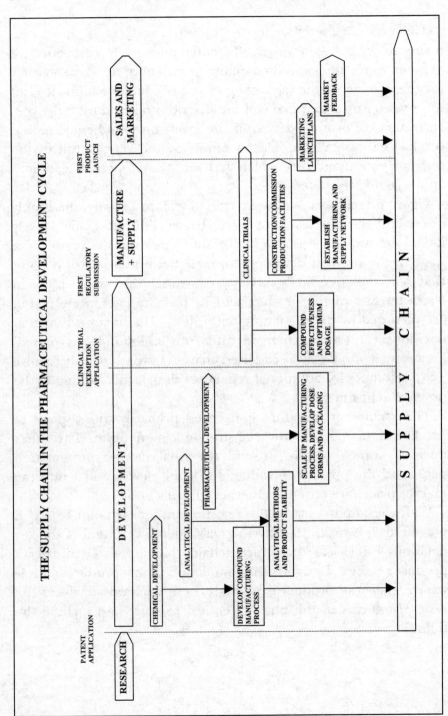

Figure 6.1

The influences of the properties of the ingredients and the primary packs on the Supply Chain and Design Parameters

Interrelationships between the Supply Chain and The Properties of Ingredients, (Formulation, Process Manufacturability, Product & Cost)

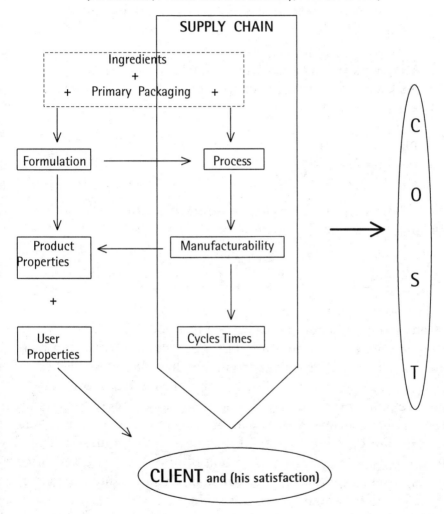

Figure 6.2

Design Parameters

- Ingredients and their choice,
- Ingredients and their specifications,
- Product (formula) and its specifications,
- Primary packing and Components,
- Analytical methods and limits for Ingredients and Products,
- Sampling regimes,
- Packaging Materials,
- Manufacturing Process and Equipment,
- Packaging Process and Equipment,
- Ease of manufacturability,
- Clinical Trials batches,
- Pilot scale trials,
- Production runs,
- Validation

All these subjects are connected with the supply chain and with Quality. Figure 6.2 shows the interrelationships.

Ingredients

The source of supply of raw materials, active ingredients, and often excipients or adjuvants, must be identified at the formulation stage. Important business commitments, contracts, and long term commercial agreements, have to be established no longer with potential, but with real suppliers, even before one can be sure that a product, or series of products can get to the market. Heavy financial investments can be involved. It is therefore vital that professional buyers should negotiate initial contracts. This is to prevent oneself being tied to one supplier, or to a fixed price, or fixed specifications. The presence of Quality Assurance, as well as production people, is also advisable.

It should be remembered that by an Ingredient, we mean anything that is used to make a pharmaceutical product. This may include items which are *not included* in the final product. For instance, processing aids such as water or alcohol, (which are required for the granulation of tablets), but which are evaporated during the drying operation.

Ingredients are very often chosen at the preformulation or formulation stage. The galenist or formulator may choose anything which is handy around his bench – just in order to see whether he can make a product. Often this ingredient is difficult to obtain, or it might be expensive, or might have specifications which are in excess of those necessary. Other items, such as lactose, used in the manufacture of tablets, can have different flow properties – depending on how it was produced. Often these characteristics are not specified, and in many instances, it is very difficult to specify. However, a large number of ingredients are perfectly "specifiable" and the sooner they are specified, the more likely it is to attain quicker, smoother, and cheaper manufacturability.

For instance the use of inflammable material (often used as solvents in tablet granulation) can have a considerable effect on the supply chain. Formulators often forget that special permits and licences have to be obtained if they are to be used in larger quantities. Of course, at the formulation stage, the quantities used are small or moderate and permits may not be necessitated. Not only are permits necessary for industrial quantities, but governments exercise special controls over these products. Special stock keeping records and accounting of losses have to be kept, thus making the chain heavier and more expensive to manage. Apart from these administrative constraints, there are safety considerations such as fire and explosion preventative measures requiring special storage conditions. It has often been found that with a little extra thought and time, inflamable solvents could be replaced by water or some other innocuous product.

If the stability batches and clinical trial materials are made using the same ingredients, then one is bound to use these materials later in production because these are the items which were registered. Subsequently, it is the supply chain that could become inefficient as a result of the original choice. Any change would require making new stability batches and validation in order to obtain a change in registration. This could take a year, and might even be refused by the authorities on the grounds of insufficient justification.

Above all, it is important when designing a product and choosing its excipients that there should be only one unique formulation per dosage form. This will greatly simplify the supply chain downstream especially in the manufacturing and registration stages. There are numerous examples of products which have been developed with slightly different excipients depending on the market in which the product is going to be sold. One cannot underestimate the benefits of standardisation for the supply chain because each variation will of course require registration. Indeed the subject is so important that it is explored in greater depth in Chapter 12 – Regulatory.

Primary Packing Materials and Components

Primary Packing refers to the material which is in immediate contact with the product. It may be the aluminium foil for tablets, glass vials for ampoules, aluminium containers and valves for aerosols, plastic or aluminium tubes for creams. They should be defined as early as possible, since they are critical for product stability and therefore implicated in the registration process.

These items have to be chosen with just as much care and attention as the ingredients in the dosage form or drug product. Firstly, because of compatibility between the product and the container, secondly, because of patient usability and thirdly, due to the manufacturability of the final product in the component. Some powder inhalation devices may have 30 components in contact with the product. All have to be proved to be innocuous to the patient and not to affect the formulation and stability of the product. It should be remembered that all of these 30 components will be part of the supply chain, when the product is to be manufactured. (see Chapter 1)

The limits and tolerances in size (dimension), physical behaviour, aesthetics and chemical constitution, have to be carefully designed and subsequently measured and controlled.

An example to illustrate the importance of this is the case of rubber stoppers used in injectable antibiotic vials. For instance, a particular new antibiotic was tried with every rubber stopper available on the market, but each antibiotic became unstable with each rubber. The stoppers contained a number of ingredients to give it the properties it required. It was found that one of these ingredients reacted with the antibiotic which caused the instability. The rubber had to be reformulated and all appeared to be well.

All went well until one tried to use it! A syringe needle had to be plunged through the rubber to add water to the antibiotic. After it had been shaken, the needle had to pierce the plug again to extract the dissolved product. The problem was that the tip of the needle cut a tiny piece of the rubber and sometimes clogged up the needle. The change in formula of the rubber altered its physical characteristics and rendered it unusable. Many months of reformulation were lost before arriving at a suitable formula.

From the supply chain point of view, this rubber stopper was a special item which the supplier had to make in relatively small quantities on a mould which was used for standard plugs. Furthermore, the rubber mixture, being unique, had also to be made specially – all this added to both the cost and the lead time.

Examples where design and design specifications are of key importance may be found with Aerosols valves and their components. Another example is provided by the components of dental syringes – (tightly fitting rubber pistons in glass tubes), where both the glass diameter and the rubber pistons have to fit tightly and yet is still able to be pushed easily by a dentist's hand.

Once again standardisation is of the utmost importance in components and primary packaging materials. Not only is the supply chain simplified but also leadtimes are reduced. An example illustrating this concerns an aerosol product which had the same formulation, but different cans for a certain market. The market that had embossed cans required an urgent supply of product from a factory which normally manufactured the product with plain cans. Embossed cans had to be specially ordered from a supplier in the market concerned, resulting in a longer leadtime than if the factory's usual supplier had been used.

Customer Preferences – Design and "Consumerism"

The customer, especially the patient who uses a medicine for long periods or, indeed, for life, looks upon his medicine or pack as part of his life. If he is then paying $ 10 to 50 a month for this, he expects and should expect something as "attractive" or patient friendly, (not to use the word "trendy"), as possible.

These remarks are especially true of the more sophisticated western societies, where a certain "consumerism" is prevalent.

Mr Hayek, the creator of the Swatch calls this "emotional consumer products". Customers become attached to what they find not only useful but attractive or aesthetically pleasing.

The pharmaceutical industry has not been particularly clever in this field, the author feels that there is much to be done to enhance packs and modes of administration and thus enhance commercial advantages.

A hypertensive patient who is not ill and does not consider himself as a patient, but is nonetheless a likely daily user of medicines, requires his pack or presentation to be not only easy to use, but also to be attractive. Asthmatic patients, diabetics, and cholesteraemics may have a similar reasoning.

It is up to the inventiveness, originality and the design sense of the manufacturer or of marketing to provide competitive presentations.

As Shakespeare wrote:

"If it be true that good wine needs no bush,....... yet to good wine they do use good bushes,". (As You Like It).

More and more use is made of "devices". For instance, for asthmatics, diabetics, etc... there is a lot of room for improvements which are designed and engineered in a cost effective manner.

The supply chain could thus well include a design function or use outside (often engineering) firms to do this and at the same time to assure the manufacturability of a pack.

However, it must be heavily insisted upon that a clever design should not be at the price of difficulty and high cost of manufacture. Often design firms are very conscious of this, and they propose very cost effective alternatives.

The quality issues are very important in both the design and the manufacture of innovative packs, presentations, and devices. Design must not be at the expense of reliability; innovation must not be at the expense of machinability. Strict quality specifications must be established and monitored. User handling tests, including ease of handling, and ease of opening packs and bottles, should be incorporated into the quality control protocols.

Analytical Methods

Analytical methods which are used to prove the purity and identity of ingredients and drug product not only have to be effective and reproducible, they also have to be registered within the Product License Application (P.L.A.) or New Drug Application (N.D.A.).

There may be different methods of arriving at the same result; yet the method registered is the one to be used.

This has significance in the supply chain because the analytical controls and the time it takes to conduct them, is part of the production cycle. Sometimes, it takes longer to analyse a batch of tablets or ointment than it takes to make it.

If a shorter analytical method exists or is subsequently developed, it cannot be used without prior re-registration. This takes time, justification, and money.

In some countries, the Certificate of Analysis (C of A) suffices as proof of the purity and identity of the material. In other countries, the company manufacturing the pharmaceutical product has to re-analyse the material and validate the analytical methods. In any case, this has to be done prior to the P.L.A.

The C of A if used, avoids the manufacturer having to retest the ingredient or component on arrival at the plant (that is if it is allowed by the local regulatory authorities). If the analytical method which the supplier uses is not the same as the one registered, than the method used has to be validated in relation to the method registered. Either for technical or financial reasons, some suppliers do not wish to validate. In this case either the manufacturer has to re-do the test – thus increasing costs and time, or he has to do the validation.

Analytical Limits

Analytical Specifications have limits. For instance, a tablet might weigh plus or minus 5 or 10 % of its declared weight. Similarly, an ingredient could be plus or minus 1 % pure. Therefore, there is a certain permitted amount of impurity allowed, provided this impurity is known and proved to be harmless to the patient.

If these limits are too tightly set then the situation might arise where the supplier's product is outside the limit of acceptability. The result is that the ingredient or component in question cannot be used, and has to be refused on arrival at the plant. This causes the supply chain to be interrupted, and the ingredient has to be re-ordered and, presumably, another lot made by the supplier. This could take a great deal of time, and could endanger the stock level which would result in an out-of-stock situation.

For many ingredients used in pharmaceutical formulations, limits have already been established by certain pharmacopeias and regulatory authorities. Some companies set their limits tighter in order to try to protect themselves from future generic companies. In so doing, they generally create themselves more cost, because either these limits are difficult to achieve, (and some batches have to be rejected), or they are not all that difficult to achieve, and then the generic company can copy them. Some companies set limits too tight just out of arrogance. The same scenario exists for finished products. There are, however, cases where this strategy does work and a competitive advantage might occur.

Another example of the importance and indeed the difficulty of setting analytical standards is that of particle size in micronised products.

Particle size analysis is a difficult and fairly complex matter. The method of analysis often depends on the instrument used, also the instrument itself can be set in different ways. The particle size can be assessed by weight and/or by volume. Different mathematical models may exist to assess distribution, averages, maxima, and minima.

Analysts look at reproducibility of results. Regulatory people look at results comparative to those obtained with clinical trial batch results. Manufacturers look at constant parameters in machine setting to obtain standard results.

The problem as regards agile and efficient manufacturability i.e. factors which affect significantly the supply chain, is that often no one looks at the whole problem. As a result, either analytical norms are difficult to obtain regularly or manufacturing parameters are not optimised. This results in batches having to be rejected, often for no good intrinsic reasons. Very often the analytical results have little bearing on drug availability to the patient.

Sometimes limits are set, for instance, for particles of a certain size in inhaled products. Whereas the important thing to know is what amount of useful particles reach the necessary areas of the lung and what quantity is absorbed by other means.

This problem exemplifies a major issue in the industry, that is the need:

a) to work together between the different functions. In this case: analysts, clinicians, regulatory, and manufacturing people and,
b) for these to keep their collective eyes on the patients or customer needs.

In the intricacies of technicalities, one can easily loose *sight* of the actual needs required by the patient!

It is therefore recommended that the setting of limits should not be done alone by either development people or registration people. It should be done by a team containing representatives from Quality Control, Quality Assurance, Production, Regulatory, and Development functions. The views of the Purchasing department should ideally also be taken into account. The test methods and limits should be realistic, capable of being validated, and user oriented.

Pilot Scale

The pilot scale process is the making of a product between the laboratory scale of between 1 to 5 kilos, and the manufacturing scale which may be from 50 kg to 5 tons. Therefore, we are talking about quantities from 5 kg to 50 kg. Generally, special pilot scale equipment is used.

The main reason for making something on the pilot scale is the lack of availability of raw material – generally the active ingredient. Pilot scale work can usually prove principles such as a spray dry granulation process for tablets, or dry mixing of powders for injection (proving that there is no demixing during the process). In the case of a cream, the pilot scale process can prove that an emulsion is stable, and there is not likely to be separation of phases.

Regulatory authorities generally recognise the suitability of pilot scale batches for registration purposes. Consequently, pilot scale stability batches can be used as a basis for establishing manufacturing processes, and also providing information on shelf life.

But pilot scale process must be viewed with some caution, because during scale up, all sorts of differences could occur. Often one finds that a product in a mass of 7 kilos behaves in a different way than a product of 700 kilos. Heat transfer coefficients are different, also total heat or drying characteristics are different. Time of contact between mixers and powder, or materials, is different. Cooling times are different and the molecules are exposed to constraints in a different manner and at different times. For these reasons, the sooner a pilot scale can be converted to full scale the better.

From the supply chain point of view, different masses (or different amounts of products) behave in different ways. In a hopper containing 500 kg of granules to be tabletted, granules flow differently and may be very poor compared to the flow in a small hopper of 5 kilos. The raw material in the hopper is generally a mixture of the active ingredient (over which the supply chain has very little control) and other materials which the supply chain may have a great deal to do with. Whether it be sugar, starch, sodium bicarbonate, or titanium oxide; – these items have to be specified, purchased, controlled, stored and dispensed.

How these products are handled, in what type of containers they are delivered, in what quantities they are delivered (perhaps in the right batch size quantities, so that no further weighing is necessitated), are all matters for the supply chain. What quality controls can be done by the supplier, and what other products the supplier manufactures, are all issues which effect the supply chain. If a supplier only makes sodium bicarbonate then there is no reason to look for magnesium carbonate as an impurity. If, on the other hand, the manufacturer of a dye makes other dyes, then more substantial controls have to be made.

All these factors are determined, implicitly or not, at the pilot scale stage of development. If they are not, then they have to be determined at the next stage i.e. the manufacturing stage by which time there might be such pressure as to allow no leeway in choice of grade of material, of supplier or of method of delivery – hence limiting price and delivery negotiations.

The variability of the physical characteristics of raw materials and excipients are very often either unknown, not appreciated, or both.

The active ingredient itself has probably been produced in small quantities and its physical properties would be different later on when the substance is made in large quantities.

Closer collaboration between synthetic chemists and formulators, could result in obtaining an active ingredient with physical characteristics which could facilitate manufacturing. For example, altering the crystal shape (habit) of a drug substance

could permit direct compression, which is much less time
consuming and cheaper than wet granulation.

Manufacturability

Ingredients, adjuvants, packaging components and their
respective specifications not only influence the properties and
quality of the final product, but could also have a major effect on
the manufacturability of a product.

Very often one comes across problems of manufacturability i.e.,
difficulty of and reproducibility of manufacture. This can be due
either to weaknesses of formulation and/or the physical
properties of excipients or primary packaging components. One
has seen on numerous occasions a certain plastic having being
chosen for its compatibility with the product, and yet could not be
properly sealed on a high speed packing machine. The original
trials were carried out on small hand operated sealing machines.
Similar experiences occurred with ointment tubes which leaked,
and bottles which broke on the tightening of the caps. Lack of
manufacturability may of course be due to the manufacturing
process, which is not the theme of this section.

The problem of the effect of poor design can best be illustrated
by the following example. Lactose is used as a diluent in the
manufacture of tablets. There are few manufacturers of this
relatively inexpensive ingredient (perhaps $ 2100 a ton).

The chemical analysis has to conform to a number of
pharmacopoeial standards. Yet there are some grades of Lactose
which have slightly different physical flow properties from others.
These properties are not part of the analytical standards, indeed
they are very difficult to measure or quantify. If the formulator, by
chance, choses a grade which is not specified, it might be later
found that another grade was used in making the tablet. One
might find that the tablet does not compress properly or it might
break or split too easily. It could take months to find the cause of
the problem. It may already be in production, and the product
may well be already on the market. The supply chain, the quality,
and the client, all suffer!

The Manufacturing Process:

It is obvious that from a supply chain and general economics point of view, the Manufacturing Process should be:

- as simple as possible,
- as short as possible,
- as inexpensive as possible,

but the most important is that it should be:

Reliable and perfectly Reproducible.

These requirements are obtained by a combination of:

- equipment,
- materials,
- methods,
- trained staff.

This is where it can be best seen that the extent of the supply chain is vast, and everything is interconnected. These factors are even more pertinent at the new product launch, where generally there is a rush. Also experience is lacking, and supply shortages of ingredients or components might exist.

An example to illustrate this point was the case of a new antibiotic powder having been launched in a novel shaped vial. After the first 100 000 vials were manufactured, it was found that too many vials broke on the filling line. So it was decided to have dummy runs to refine the performance of the line. After a further 50 000, the problem considerably diminished. However the 50 000 was unforeseen and a very near out-of-stock situation resulted. Because it was a special vial, the glass manufacturer had to replan his moulding production – not an easy matter.

Another example is the case of a tablet which was coated with an alcoholic solution. When stored for the first time in large amounts, it created a concentrate of alcohol vapour in the bulk container which was potentially a fire hazard. Only the careful observation by smelling, avoided not only a supply problem but also a hazardous safety situation.

More is said about manufacture in Chapter 8 of this book.

The Packaging Process

The supply chain problems which are the most annoying with the packaging process concern the lead times of packaging items such as aluminium foil, plastic foil, moulded bottles and vials, and moulded plastic items.

Printed materials such as labels, leaflets, and cartons, can easily be modified and manufactured anew.

One has often seen that foil produced in industrial, as against experimental quantities, show different sealing and hermetic properties. This endangers the quality of packaged items such as humidity sensitive tablets or capsules.

Bottles with ill fitting caps can create equally serious problems, which are often found after large scale and high speed production or packaging operations.

The ideal way to avoid these unforeseen problems is again to have two approved suppliers – which is not always good economics or even possible at the time of a product launch. Again this point proves that the supply chain starts at the design stage where two suppliers should be identified, and technical terms of supply established.

Clinical Trials

Clinical Trial material is generally manufactured and packed on small scale equipment, thus little reference or comparison can be made with future large scale production. However, careful observation as to how a product behaves under even a semi-industrial handling condition can be very important. Minor observation as to sealing of plastics, flow of tablets, flow of liquids and ointments can avoid major problems later and can also influence the choice of supplier.

The problem is that very often clinical trials material is prepared by a separate unit, that is, separate from both production and purchasing/logistics people. They are usually not trained to report anomalies which could be relevant in the future. It is therefore highly desirable to have at least one person in the team who has production/supplies knowledge so that feedback can take place.

On the other hand, there is also the danger that if all runs well during a clinical trials manufacturing and packing exercise, one might think that all will be well during production runs (see importance of Pilot Scale manufacture and scaling up).

Fitting caps on bottles and tubes is a classic example where slow hand operations might be easy; but at the large scale packing stage, caps or tubes may split. The temperature and setting time of emulsions is another example where the unexpected could arise, due to the scale and speed of operations.

Validation

The main purpose of validation is to prove that the manufacturing and packing processes are reliable and reproducible. Speed of the process is generally not a factor in this operation, although of course critical times must be adhered to.

The fact that speed is not a consideration at the validation stage could present surprises at full manufacturing production levels. This is often forgotten!

When, later on, the manufacturing process is optimised, it must be done in such a manner as not to be in contradiction with the data gathered during the validation exercise.

Generally, it is not the critical parameters such as speed and the length of time in the stirring of a syrup or cream which influence the speed of manufacture. The time it takes to manufacture is much more influenced by the sequential nature of the process, such as addition of powders to a granulate or the addition of flavours after drying of powders. These are steps in a process, where machines have to be stopped, where the product may have to be transferred from one vessel to another, or a granule may have to be resieved. When properly optimised, it is often found that these steps can be avoided, allowing certain products to be added concurrently with other ingredients.

More imaginative use of equipment, more experimentation at the time of validation, could well lead to the shortening and simplification of the manufacturing cycle (see Multibatching on page 110 of this book).

Expression of Needs

In many cases the user or manufacturer does not know how to express his needs. Very often, he is not trained for it. A pharmacist tends to concentrate on the intrinsic quality of the product rather than "what goes round it" whether it be excipient, primary or secondary packing material, etc...

As can be seen from what has been mentioned before, very often quality problems are unnecessarily created by registering specification to the authorities which are too tight for finished goods. Suppliers, or the manufacturer himself, might at times be unable to comply with these. And although both the product and the pack may be perfectly usable, and the client perfectly satisfied – the product cannot be made, or is sometimes rendered unsaleable purely for legal reasons.

Development as stakeholders

In summary, the important point to be highlighted in this chapter concerning the development of new products is that the supply chain must be established as early in the process as is feasible. Thereby involving as many of the downstream interested parties as possible (manufacturing, industrialisation, logistics, quality assurance, control etc). All these personnel should have at least some consultative input into the design of the product before design parameters are fixed by regulatory submissions. When design parameters are specified, standardisation should be the name of the game. In this way one is able to develop a product in the shortest possible timescale with a supply chain that has been optimised for the most important stage of a product's lifecycle – its time on the market!

Development must view themselves as stakeholders in the supply chain, and be held accountable for its success or shortcomings. This is what Empowerment and the Learning Organisation is all about!

Critical factors:

This part concerns purchased in items i.e. not those manufactured or produced in-house.

- Early involvement of supplier and user. This means not theoretical ivory tower specifications.
- Simultaneous development,
- Mutually set-up and agreed specifications,
- Clear specifications,
- Strict Modification procedures,
- Balanced and Pragmatic Quality Assurance involvement,
- Machinability,
- Flowability of powders,
- Cost – benefit assessment. Procedures and Records.

Quality Actions:

- Programmed updating and review of specifications,
- Auditing of specifications,
- Automatic systematised and fast feed back of user problems
- Early manufacturability trials if necessary,
- Close and critical monitoring of Pilot plant results.

Critical Quality Indicators:

- Number of reformulations,
- Time for prototype to Clinical Trial Research,
- Time for prototype to Market,
- Number of prototypes to Market.

7 | COMPONENTS

This concerns all bought in items such as excipients, packing materials, containers, etc...

The particularity of components is that:
a) sometimes they are cheap – inexpensive at least compared with active materials,
b) quantities ordered might be small even for the supplier – such as for example: flavours, colouring materials, and rubber components.

This could result in the supplier being unable to afford to be as cooperative and flexible as one would wish.

If his minimum order quantities are well in excess of requirements, it is advisable to hold larger stock which, if this stock is inexpensive, will not present a financial burden on the supply chain.

A further fact, often associated with the above, is that the supplier manufactures the item only once or twice a year. Thus if there is an urgent need for larger quantities than usual, or there is not enough available at the supplier, the lead time could engender major supply problems. The user might be tempted to seek another supplier whose product has not been validated and tested according to the usual criteria. This could result in a major quality problem.

The only issue then might be to choose suppliers who have not had previous experience in supplying the pharmaceutical industry. These have not only to be carefully trained, but also strictly monitored.

The technological capabilities existing within a supplier must be known, as well as their own research and development activity, to improve their process, quality, and their capability to innovate.

Sharing with a supplier one's own technological perspectives is also essential. If for instance, a company is going to adopt On Line Printing, then the supplier who has previously supplied printed cartons must be made aware as quickly as possible that he will only be required to supply plain cartons. Thus his investment programmes could be highly affected.

Printed items present other potential problems. Exactitude and precision in texts is vital in the pharmaceutical industry.

The supplier might change a specification which he thinks is minor, but could nevertheless be very important to the manufacturer. For example, physical characteristics of Sodium Citrate used in tablet manufacture.

It is crucial that the supplier be aware of the vital consequences of the quality of *his* suppliers on the final consumer. This awareness must be spread throughout his organisation, and must not stay only at the commercial level.

Any changes, either for consumer or regulatory reasons, have to be approved not only inside the organisation by marketing and medical, but generally also by legal authorities. Indeed, in most countries, commercial publicity/texts have to have government approval.

A comma, or a letter in a word, being omitted can imply legal infringement. Therefore texts have to be created, approved, and checked by competent services.

Any changes or alterations have to be approved by the respective departments and the Quality Assurance of the firm.

It is vital that this approval and checking process be as rapid and as transparent as possible. Even fax transmissions are inadequate to assure precision. A typical error can show us a measure of the complexity of the problem: *5 pour 100* i.e. 5 % if the *pour* lacks the *r*, i.e. it is pou – result unconformity. *pr*, is it OK.? *p* is it O.K.? This type of problem can cause delays of up to 2 weeks which can put back an order worth $500.000, or sometimes lose it altogether.

Undue pressure may be forced on Quality Control to sentence materials. It could take unnecessary or unwarranted risks resulting in vast consequential risks on quality.

Electronic liaison with suppliers and new technologies, such as "print to plate" would attenuate some of these problems. In this case, the printing plate is etched directly from the "proof" thus avoiding human intervention.

Storage of Components

The storage conditions of components are just as important as those for active raw materials and finished goods.

Cartons and labels, if stored in adverse conditions of temperature and moisture, can have disastrous effects on the supply chain. High speed machines breakdown if cartons are too moist, and one may think that one has a stock of 150 000 cartons, when in fact it is zero, as they are unusable.

Excipients can be just as sensitive to moisture as active ingredients, and the same care in their bulk packaging and storing is required.

Often suppliers want to cut corners in cost terms by using cheaper bulk containers with ill fitting or poor closures. This is an expensive practice.

Protection from vermin must be equally efficient.

Generally when one's main warehouse becomes full, one looks for outside storage space; this may be an inflatable tent or just a plain tent. It might be another building of one's main site. Simplistically one may think to store the cheaper components in such alternatives.

This course of action has to be very carefully assessed because the protection given to components must be, as said above, adequate.

In fact, any storage conditions should be properly validated; that is, the conditions of temperature and humidity known throughout every period of the year. Likely maximum and minimum conditions should also be known.

A simple example of a well known problem is the case of glass bottles and aluminium cans which are stored in non temperature controlled stores. If these items are at a temperature of 3 or 4°C and then they are brought into a filling room which is at 20°C, condensation will form inside the containers. If the product, which is to be filled into the container is moisture sensitive, then major quality problems could arise.

Generally, in these circumstances, the items are brought into an area which is warmer – may be 10 to 12°C and kept for a few days prior to use.

Transport and Handling of Components

Similar precautions apply to transport and handling arrangements as for storage conditions. It must again be emphasised that even if an item is relatively cheap, it must be considered as important and sensitive in the same way as an active ingredient which may cost one hundred or one thousand times more.

Just-in-Time Delivery

The introduction of Just-in-Time delivery leads to careful re-definition of the way goods are packed for delivery, the way they are stored and transported.

This reconfiguration must be done in close collaboration with the supplier, his transporter, and the receiving company or site.

Critical factors:

- Advanced technological awareness,
- Likely future evolution,
- Awareness of lead times,
- Awareness of minimum order levels,
- Awareness of frequence of manufacture,
- Unequivocal specifications and their comprehension and validation (sample) by supplier,
- Back up, or two regular suppliers,
- Speed, precision and transmission of artwork,
- Artwork approval system,
- Market pack approval system (if relevant),
- Machinability of components,
- Usage value (expected reject rate),
- Cost of investment for further capacity,
- Cost of investment for replacement capacity,
- Storage conditions,
- Research and Development capability of Supplier
- Quality organisation of supplier (ISO or other standard)
- Safety systems and procedures.

Quality Actions:

- Regular supplier audit,
- Programmed updating and review of specifications,
- Auditing of specifications,
- Automatic, systematized and fast feed back of problems,
- Improvement programme agreed with supplier,
- Auditing of supplier's maintenance schedule,
- Certification programme.

Critical Quality Indicators:

- Quality Control acceptance/refusal trend analysis,
- Reject rates: – at receipt,
                              – on line.
- Delivery times to replace faulty items,
- Storage condition records,
- Social Climate within supplier,
- Safety records.

8 MANUFACTURING

This is the most "controllable" part of the supply chain because it is in-house. (We are not talking here about subcontracted or outsourced manufacture). In fact, it is in microcosm; the whole subject of this book, starting from the supplier and ending with the user.

The Supply Chain in Manufacturing

By a better and fuller understanding of the mechanism of the Manufacturing Process within the Supply Chain:

- one can better visualize the flow of goods,
- one can assess the cost of the flow,
- one can analyse the constraints which impede the flow,
- one can understand the quality impact on flow,
- one can understand other factors affecting the flow,

all these, in turn, can therefore lead to a more cost effective and better organisation of the chain.

Continuous Improvement and Agile Manufacturing

Continuous improvement is one of the key actions within manufacturing. This is the road to agile manufacturing, which in its turn will lead to mass customisation. It is being realised that the advantages of economies of scale has limitations. Cost can only be reduced and efficiency increased by making more of the same thing. Costs go up if fast speed equipment is used for smaller runs. Smaller production runs for specific products and/or markets, require rapid change overs, flexible equipment

and even flexible and easily adaptable work areas. S.M.E.D. *
and O.E.E. ** are tools which are essential to reduce cycle times
but responding to unexpected demands requires anticipative
actions and agile management techniques. These include
immediate demand feedback, smaller equipment, multipurpose
equipment, multiskilled and mobile staff, etc...

Understanding the processes:

Manufacturing processes are not simple. Unless one has
personal experience, understanding requires quite a lot of
explanation and observation. (This is perhaps why some of the
best logisticians and supply chain managers have previous
production experience).

The first problem is to understand the sequence of the steps. It
is obvious that a tablet cannot be coated before it is compressed;
but why does a tablet need to be compressed? and what will
happen if it is not compressed hard enough? (It breaks as it is
packed, or as it is transported – thus affecting the quality of the
supply chain).

Why does a tablet have to be coated? Is it to hide its taste or to
make it run smoother and faster on a packing machine? Certainly
the latter would shorten the supply chain but it is probably coated
for both reasons.

Secondly, it is necessary to know the length of time of the
sequences and the reason for the length. Why, for instance, does it

---

* S.M.E.D.: Single Minute Exchange of Die.

A productivity improvement initiative aimed at reducing format changeover times
and thereby reducing and/or eliminating non-productive time. This initiative is
usually applied to production lines but could be equally applied to any type of operation where two tasks or operations are interrupted by an obligatory non-productive
period.

** O.E.E.: OVERALL EQUIPMENT EFFICIENCY

A measure of productivity which compares the amount actually produced by a production line (or piece of equipment) with a theoretical maximum achievable at a specified nominal speed during a defined working period. The nominal speed should be
the speed at which the line is set to run under normal conditions and the defined
working period could be anything from a single day shift to a round-the-clock, 7-day-a-
week operation.

take 4 hours for a coated tablet to dry or mature before it can be blister packed? Could it be dried faster by using higher temperature, or would its quality be adversely affected?

The programming or scheduling can be done in different ways; it can be pushed or pulled, (see section on Kanban). This can have considerable influence on the supply chain and thus the quality of the product.

Constraints related to Manufacturing

By understanding the production process, one can make a list of constraints to be eliminated to improve the supply chain. This is necessary in order to reduce cycle times without adversely affecting the quality of the product.

Although it may be tedious for some, it is still worthwhile to run through a typical pharmaceutical production cycle.

Figure 8.1, shows the work flow.

At each of these points and between each of these points time elapses.

At each of these points constraints could occur. It is the thorough knowledge of these times and their constraints, *by all* the actors along the chain, that will help to shorten the chain, i.e. reduce and simplify the cycle time.

Figure 8.1, shows also a typical chain time, generally referred to as "Internal Delay".

Waiting Time:

As one might have noticed, most of the time spent is "waiting time".

This means:
a) the raw material or packing material *waits* for sampling,
b) or it *waits* for testing by the Quality Control Laboratory,
c) it *waits* to be allocated,

d) it *waits* to be weighed, etc...

Thus 85 % of the cycle is *waiting* time.

Waiting is "non value adding". In fact, according to the Value Flow – it is actually losing value (see Introduction, page XXIII) which means that five days are lost.

The actual manufacturing time could be one day. But as can be seen from the diagrams, the cycle actually takes 6 days.

Let us analyse this.

Continuous production process is not typical in the pharmaceutical industry, batch manufacture being the usual practice. A batch, or series of batches, of a product is programmed by the Planning or Logistics department. This is done mostly on a campaign basis, where a number of batches to be made on the same equipment, or series of equipment, is grouped together. This is in order to reduce cleaning time between different products, and also to reduce down time of equipment. (Changing of punches and dies can take 3 hours! Changing from one size of carton to another on a packing line can take 5 hours).

Generally, these campaigns are optimized as a function of:
– quantitative requirements,
– capacity of equipment,
– equipment scheduling,
– availability of staff,
– multiskilling of staff (cross trained workers),
– length of cycle,
– time available,
– shift arrangement,
– cleaning times,
– machine down times.

While each batch is processed, the other batches of the campaign are prepared. They may be pre-weighed or take part in an early part of the cycle.

Hence they wait.

A similar situation occurs when ten cars wait – or queue to cross a bridge – when the bridge can only take one car at a time.

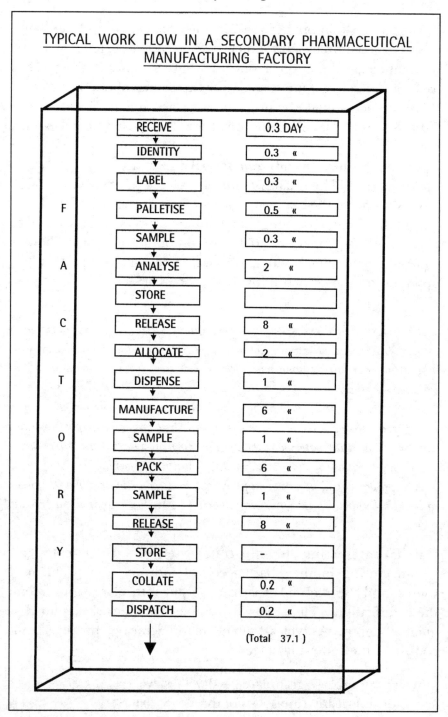

TYPICAL WORK FLOW IN A SECONDARY PHARMACEUTICAL
MANUFACTURING FACTORY

	RECEIVE	0.3 DAY
	IDENTITY	0.3 «
	LABEL	0.3 «
F	PALLETISE	0.5 «
	SAMPLE	0.3 «
A	ANALYSE	2 «
	STORE	
C	RELEASE	8 «
	ALLOCATE	2 «
T	DISPENSE	1 «
	MANUFACTURE	6 «
O	SAMPLE	1 «
	PACK	6 «
R	SAMPLE	1 «
	RELEASE	8 «
Y	STORE	–
	COLLATE	0.2 «
	DISPATCH	0.2 «

(Total 37.1)

Figure 8.1

The same situation can arise in the Quality Control laboratory where it is simpler and quicker to test five batches, virtually simultaneously, than to test one at a time. Samples of five batches, each of which is manufactured over the course of one or two days, therefore accumulate before the tests start thus making the cycle for the tests 12 days! During this time, the finished batches must wait.

Rarely is the *cost of waiting* balanced with the so called optimization of the Quality Control cycle or the process cycle.

This is one of the major problems of the Internal Supply Chain; unless the factory is using the Kanban process or has been re-engineered in a manner to favour or optimize process flow. It contrasts with optimisation programmed in the past, where optimisation concerned individual process stages.

There are many examples of where queueing takes place; items to be ordered from a supplier are generally accumulated before the order goes off. The supplier similarly often waits for all his items before he sends them off, grouped together.

Similarly, Accounts Department will accumulate a number of invoices and process them, perhaps, once a month. Mail is generally posted once a day, meaning that letters written in the morning could wait for 6 hours.

All these examples show the "waiting" nature of campaign work.

A supplier would be much happier not to have to wait a month for his bills. His Supply Chain only ends when he gets the money for his goods!

It is interesting to illustrate the effects of campaign manufacturing in the pharmaceutical industry, the problems it causes, and the methods that can be put into place to overcome these difficulties. The following simple example is examined in greater depth. A typical tablet manufacturing process could consist of the following stages :

Weigh → mix → granulate → dry → sieve
→ add lubricant (mix) → compress → coat

Equipment scheduling and efficiency:

All these stages are effected on equipment or machines. Thus each is a "mini" process itself. Each step may take from 1 to 15 hours. Therefore, it has to be planned or scheduled. (Each may be considered as cars crossing a bridge one at a time).

Cleaning process:

The cleaning process between batches (even of the same product) can itself take from between one to four hours. However if there is a change of product – then the cleaning time could take 10 to 50 hours.

Let us say that a batch of 500 kg of tablet A from weighing through to mixing (prior to compression) takes 14 hours to process. Because of this, we may make five batches in a row in order to avoid major clean-downs.

So, for approximately five days, the weighing, mixing, granulating, and drying equipment is dedicated for product A.

At the end of these five days, we have enough product for, let us say, two and a half months' sales.

This is the value of the stock we have to hold.

We can then clean, which let us say, takes two days (18 hours work). Next we can proceed with making 1 batch of product B, of which we need much less. After this, further cleaning takes place in order to make either product C or another batch of product A.

As said before, the process is essentially sequential. Provided there is enough staff available, one person can weigh whilst another granulates.

The problem of campaigns is on the one hand, that in terms of work flow and in terms of capacity utilisation, it is better to make one batch of the same product after another. On the other hand, no other product requiring the same equipment can be made and therefore its stock level could fall dangerously.

In parallel, the stock of product A, having increased, takes up not only space but also financial resources. If a batch is worth $ 25 000 and is "lying idle" for a month and should this happen six times a year at 5 % interest rates – this costs the company $ 7 500 a year. Sometimes, the value of a batch is much higher, and we could talk about ten to fifteen products.

One can soon see the extra cost of manufacturing which, in a sense, is wasted.

Increasing capacity is the simple but expensive answer. Shift work is another answer. Sometimes improving the process and reducing waiting time and/or drying time can similarly help the supply chain.

Often one of the steps in the process creates a bottleneck. For instance, the sieving. In this case, a second or even third sieve, could give the answer.

Or perhaps a different sieving process might also speed up the process.

Multibatching (Piggy backing)

Another way to optimise production, in terms of time, financial, and capacity resources, is to combine batches. This avoids cleaning down times between batches, which in some cases can take 50 % as long as making a batch. This also reduces both analytical testing times and costs, as well as administrative procedures. The risk of course, against which this method has to be assessed, is that if there is a quality problem with one of the batches, then both batches might have to be rejected. The consequence is the doubling of financial losses.

This method has to be validated and approved by Quality Assurance. Regulatory Authorities generally allow this, provided that no more than 5 to 10 % of the remains of one batch is mixed with the next one.

Multibatching can be done for a number of batches depending on the result of the risk-benefit analysis.

This procedure has been in use for many years. In fact, it started with sterile antibiotic powders which were delivered in relatively small canisters of between 2 and 5 kilos. With large throughputs on fast filling machines, it was found that there was a greater risk in cleaning down and sterilising between each 5 kg batch, than by changing the canister in a controlled manner every 2 or 3 hours. Subsequently, a week's production was used as a "batch". The process if validated, was reliable.

(see paragraph on Parametric Release in Chapter 9).

Staff and Time

Competent and fully trained staff or multiskilled (cross trained) staff, in the right numbers have to be available to achieve the planned production. There must also be sufficient time available to complete the tasks. (One cannot leave half a batch of an aerosol unfilled over a week-end – some of it would evaporate and/or the active ingredient would deposit).

All these factors are so integrated with quality that it would be both tedious and superfluous to mention the fact at every point.

It is at this point that it is worth introducing the Kanban system and the impact that it can have on the supply chain.

The Kanban system

Kanban, loosely translated, in Japanese means *a card*. The concept consists of cards indicating a number of constants circulating within a work place. Its object is to "pull" an activity, product, or item, to its next stage. Because the Kanban system pulls the product through batch by batch, as and when it is needed, so theoretically, it avoids waiting times. Essentially, it is a re-ordering system.

Looked at in another manner, it is basically a signalling system which informs the "previous step" of a sequence of a process or operation to happen:

– What it wants,
– How much it wants,
– When it wants it,
– Where it wants it,

It is therefore said to be *pulling* the product or process.

Taking the car and bridge as a model, it is a sign informing the car that it should cross the bridge. It is the light turning green, so to speak.

The "what" is always the same for a particular Kanban card. It might be a raw material to be weighed or processed, a batch of product to be packed, or a batch of cartons, etc.

The "how much" is also always the same. It is a quantity which is to be used or processed in a *known time*. The Kanban quantities are a function of the capacity of the equipment, size of a holding vessel, size of a batch, a definite number of units of finished product, or whatever the Kanban is referring to.

As said before, there are two basic ways of planning production; pushing or pulling. When one pushes one accumulates the product or stocks between stages. When one pulls there is no accumulation or waiting time, hence theoretically, the cycle time is reduced. In practice it is rare to find a piece of equipment lying idle just waiting for the next order to come through. So one has to take into account a number of constraints which do not allow one to produce at the exact moment when it is required.

For instance, the equipment might have broken down, or there might be insufficient staff to man it, or it might simply be being used to produce another product. All these constraints have to be taken into account which is why many Kanban systems operate with some stock, the level of which is agreed beforehand, and reflects the need to include some flexibility to allow for the above mentioned constraints.

The next step is to remove or minimise these constraints in order to reduce one's cycle time, and subsequently one's level of stock.

INCREASING CYCLE TIME EFFICIENCY

Broadly speaking the method involves firstly identifying the constraints, next they have to be analysed, and then be prioritised into an action plan for their resolution.

Identifying constraints

The constraints may be of the following broad nature:
- machine reliability,
- machine capacity,
- organisation and planning,
- batch size,
- campaign,
- cleaning,
- labour (availability, training),
- administrative (documentation).

An example will be taken to illustrate this point:

Machine reliability

In a particular production cycle, 20 different machines may be involved. By machines, we mean equipment where there are moving parts, electric, or electronic sensors, etc...

balance 1

```
                  --> transferring --> sieve --> holding --> valve --> granulator
balance 2             tubes                      vessel              granulating
                                                                     vessel + mixer

 <---------------------------------------------------------------------------
 |
 |--> sieve --> holding vessel --> tabletting machine --> metal detector
 <---------------------------------------------------------------------
 |
 |--> coating vessel
    --> coating solution vessel + mixer       --> transfer mechanims
 <----------------------------------------------------------------
 |
 |--> packing machine --> code reader --> carton grouping machine.
```

Any hitch or breakdown in any one of this chain of machines can interrupt the chain during the time taken for the machine to be repaired.

In one case, a total of 5 hours delay was induced by different machines in a 45 hour cycle. Obviously, by reducing this to 2 in a constant effort to eliminate waste, the saving was estimated at $25,000 per year. Furthermore, the capacity was increased by 8 %.

Quality also improved, for breakdown in equipment creates not only product losses but also could affect the product by staying in one place far too long. Thus becoming overheated in strip packing machines or ovens, etc...

Capacity

Let us say there is a problem with a valve of a large vessel or container.

In order to repair it, the product has to be transferred to another container. If there is not one available – then there is obviously a capacity problem.

Organisation & Planning

If there is a batch, or part of a batch of tablets which is required to be packed in a special carton having a leaflet which is not standard and cannot be done by automatic machine it must be hand packed. Therefore, the product has to be taken off the automatic production line. This only serves to disorganize the packing schedule.

Batch size

If there is a special small order for a particular country, requiring a special formulation which is not the size of a usual batch – the equipment is nonetheless utilised and immobilised for half the usual quantities.

Campaign:

Making up a granulating solution takes perhaps 1 hour regardless of whether it is for 1 batch or 5 batches. If a campaign of 5 batches is made then one saves 4 hours by comparison with making 5 single batches.

The same thing happens when preparing flavours for 1 or many batches.

Cleaning:

It has been found that cleaning every day for 2 hours is more effective than cleaning once a week for 12 hours. Furthermore, during an intensive programme of validation, certain especially sensitive points in the equipment, (which needed more careful cleaning) were found. Therefore, not only was the cycle time reduced, but the quality was also improved.

Labour:

a) *Organisation:*

Operator absenteeism, waiting for temporary staff, and insufficient training of temporary staff, can all increase operating times and affect quality.

By working through tea and lunch breaks, and operating machines during these times, at least 1 hour or perhaps 1 hour and a half can be saved. This can be done by staggering breaks and lunch hours, provided of course, that the staff have been properly trained. This can reduce cycle times by 16 per cent or more.

b) *Training:*

The efficiency of a factory and its production depend on its staff, the way they are motivated and trained. Staff operating equipment can double the output if well trained. Trained not only in running the machine but also repairing minor breakdowns. T.P.M. programmes (Total Preventive Maintenance) can be run by production staff. In-process control and quality control activities can similarly be performed by production staff.

Administrative:

Experience has shown that by reducing double signatures to the legal minimum – one can gain 20 x 3 minutes of operators' time per day, without endangering product quality.

By checking paper work during the run of a batch, one can save time at the end of the batch run when there is generally a rush to get on with the next batch. This could again save up to 4 % of the cycle time.

Other Constraints

Lack of standardisation

One of the major constraints which has to be challenged concerns the lack of standardisation of products and packs. Considerable economies can be made and the supply chain significantly shortened and simplified, if markets accepted the same product packed either in a standard manner or very similar in shape and size to those produced for other markets. This is a vital element in reaching agility in manufacturing. How far this can be achieved in mass customisation, which is partly in contradiction of this, is one of the challenges facing the industry.

Fortunately multilingual packs are becoming more and more accepted in the industry. Nevertheless, we very often find that the same product is packed either in different primary packs and/or different secondary packs for different markets.

The situation is often even more complicated. Tablets might have different shapes, sizes, and colours. Syrups might have different colours and tastes. Aerosols might have a different number of doses per can.

Plastic packs for different markets could have different colours and have different texts embossed on them. Formulae might be slightly different, either for regulatory, historical, or commercial reasons.

The client, who is the patient, should get what he requires. Very often, one wonders whether he minds if the tablet is blue or pink. Similarly, when Marketing wish that the syrup should be blue, although in most other countries, it is colourless – one wonders whether the patient really cares.

Examples of economies obtained through standardisation are numerous.

By standardisation of punches and dies on a tableting machine, cycle times have been reduced from 7 h 30 to 4 h 30. We often find that the same product with the same strength and colour have slightly different punches and dies. The reason being that either the marking differs, the concavity is different, or one may have a breakline and the other not. This is due to insufficient attention having been paid at the development stage, with Marketing of each country being asked what they wanted and their wishes being fulfilled.

Another example to show the efficiencies or inefficiencies created by multilanguage packs is shown by an aerosol line.

Figure 8.2 shows an 80 % efficiency on a single product without change of labels.

Figure 8.3 shows the efficiency dropping to 52 % on a similar line producing for three countries, namely France, Belgium, and Luxemburg.

All three incidentally are French speaking but nonetheless require three different labels. On a 5 million yearly production, the extra cost represent about $500.000. This is bottom line cost and reduction of margins.

Figure 8.4 shows the difference in down-times due to various factors between a single and a multiproduct line.

It can be seen that line changes and cleaning times increase very significantly on a multiproduct line.

A further great advantage of standardisation is that of quality. Not that quality would improve, but the danger of mix up would be reduced. Line clearances would be less frequent.

Quality Control Laboratory Constraints:

Constraints exist equally in the Quality Control laboratory. Their analysis can be carried out in a similar manner as in the Production Cycle. As it is a process, the study is in fact carried out better in conjunction with the production cycle. The Quality Control activity comes in before, during, and after manufacture and can then be considered as part of the cycle.

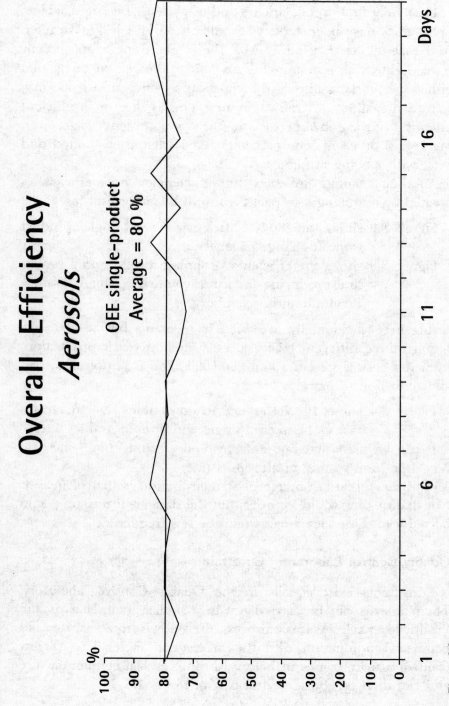

Figure 8.2

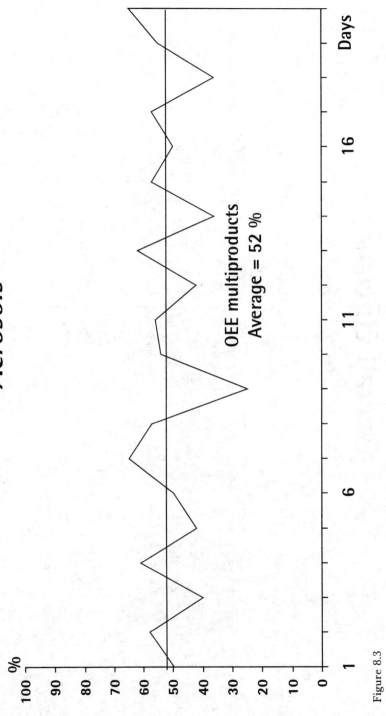

Figure 8.3

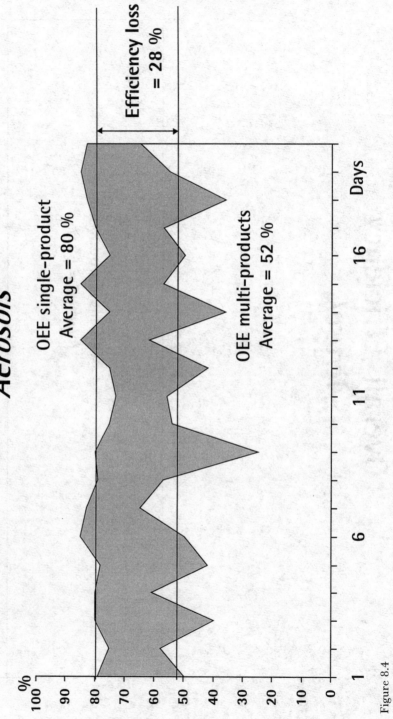

Figure 8.4

A typical reduction in analytical cycle time is the case of a tablet which took 1 hour to dissolve completely. The active ingredient, which was the object of the analysis, dissolved in 5 minutes, the excipient took one hour. By centrifuging off the non dissolved excipient – the sample could be analysed after five minutes.

Another example: chromatographic plates took 12 hours to dry; therefore, the analysis could not be performed on a Friday. By changing to another method of analysis, the same objective was achieved in 1 hour. This required re-registration but the method was more precise – quality and time went hand in hand.

Further discussion of the control of quality is presented in Chapter 9.

Analysis of Constraints:

The analysis of constraints is a long and careful process.

The most difficult problem is to persuade people to think and talk about the constraints within a process.

With experience and by asking the right questions such as:
– how long does it take?
– why does it have to be done?
– what would happen if it were not done?
– how could it be done differently?

one can generally establish the list of constraints and their relative values in time and money, their causes, and the methods of overcoming them.

Eliminating constraints is a mind set i.e. where the accepted method and thinking is continually questioned.

If we take as an example the Antilope system put in by Glaxo Wellcome in France, we can see the type of methodology which can give reduction of cycle times of 30 per cent or more.

In this system:
1) values are given to the frequencies of constraint occurrences,
2) times values are given to the constraints, i.e. for instance, if a machine breaks down, what is the average time of the stoppage?

3) How and by whom is the constraint brought to light? Is it observed by the operator or by the Quality Control laboratory or possibly even by the client?

These values are then put onto a resource/feasabilility/cost table which has the following criteria:
- cost of investment,
- technical feasibility,
- internal resource costs,
- regulatory involvement,
- length of resolution.

Each constraint is then weighted with a 1 to 5 difficulty factor. Thus prioritization is achieved in terms of:

- time,
- cost,
- difficulty,

in relation to gains in terms of cycle time and/or quality improvements.

The actions are called "levers" and the priority is given by "accessibility".

Figure 8.5, shows a typical table obtained. This table forms the basis of an action plan to eliminate or reduce the constraints.

Lead times in Manufacturing:

Different companies use different definitions for lead times.

Definition n° 1:
"Lead time is the time taken between placing the order for the raw materials and components, to delivering that raw material as finished goods to the warehouse or to the customer's transporter."

Definition n° 2:
"Lead time is the time taken between placing the order for the raw materials and components, to selling the last box of medicine made from that raw material to the final customer."

Definition n° 3:
"Lead time is any process from *beginning to end*."

CLASSIFICATION OF « LEVERS »

Constraint		LEVER	Stake lever	Stake	COST	FEASABILITY	PROCEDURE	RESOURCES	PERIOD	Accessibility
Ranitidine hopper campaign effect	50	only one code	4	4	0	0	4	1	1	18
«	50	automatic draining tank	4	4	3	2	2	2	2	25
«	50	automatic draining tank maintaining hopper	4	4	2	2	2	2	2	20
«	50	pre-weighed quantity sent by the supplier (Supply Chain)	4	4	2	3	4	1	3	37

Figure 8.5

Definition n° 4 :

"Lead time is the time between a raw material and component arriving in a warehouse and being *available for sale* as finished goods."

To each of these definitions one can add the delivery time (Dock Time).

Each definition is good, provided that everybody along the supply chain:

a) knows it,

b) understands it.

This may be obvious, but if we take an example by saying that the laboratory's lead time is 6 days – what does that mean? Does it mean that once the laboratory has the sample, it gives the result in 6 days, or that it takes 6 days to analyse the sample? Or does it mean that, on average, it takes 6 days to sample-test and give the result?

Again there is neither a good nor bad answer – but there must be an agreed *answer* – that is all.

Lead Times – one explanation:

One way of visualising the problem of Lead Times is to imagine a scenario where the raw materials and ingredients come in day 1. They are analysed days 2 + 3, and become available for production day 3. The product is made on days 3, 4, and 5, analysed on days 6 and 7; becomes available for packing on day 8, and is packed on days 9 and 10.

Lead Time is therefore 10 days – not a remarkable, but a reasonable performance. But, and here comes the but, if we have more than one or two different products or packs from the same raw materials, then this delay might increase five to ten fold. Let us say that we are talking about tablets of the same formula for different markets. This might mean not only different packs but different tablets shapes, sizes, or configurations.

We could find twenty different orders for this product, the first could be produced on day 10 and therefore the lead time is 10 days, but the 10th or 20th order could take 3 or 4 weeks! For the same product! These are some of the typical supply chain problems which occur daily in a pharmaceutical factory.

It is very easy to say that standardisation of tablet shapes, sizes, and configurations is the answer; or that standardisation and multilanguage packs avoid this situation. But the realities of the market place have to be met – that is the role of Manufacturing.

This example also shows the need for flexibility, which is one of the vital attributions and necessities of modern agile manufacturing.

A flexible and agile factory either has multiple equipment to cope in parallel with a number of orders, or has rapid change-over times – but preferably both!

It is obvious that in good supply chain management and practice, the lead time should be as short as possible. But it is much more important that the lead time promised should be kept, and that it should not vary from one order to the next.

A variability in lead time is a very good indicator which shows up not only technical and logistical problems, but also quality problems. A faulty manufacturing procedure, a faulty analytical procedure, a retest, or indeed a batch reject, generally has a negative effect on lead time.

New Product Launches

A particular problem which must not be forgotten is New Product Introductions (N.P.I.). (See also sections in chapters 2 & 5). In many cases, this is one of the main reasons for which a company owns its factory.

By its very nature, volumes are not easily forecast with accuracy. Therefore the efficiency of the supply chain is affected. These risks are compounded by the learning curve both of the production process and other business processes associated with a new product.

Therefore, it is advisable to have buffer stocks and less lean product flow schedules and systems. This is not easy to put in place because it is in contradiction to the principles of Good Supply Chain Practices (G.S.C.P.).

It could also confuse staff at both the operational and planning levels.

Nonetheless, it is important to differentiate New Product Introductions from current production, especially if there are a number of different forms, packs, and presentations involved. Once however agile manufacturing principles are in place, the differentiation is less notable.

Temporary Staff

Temporary Staff, generally in the packing operations but also sometimes in the manufacturing operations, are called upon to help when sudden surges in production occur.

This might be due to unforeseen sales, poor planning, or a supply chain problem such as late delivery of goods. It could also be due to production problems where a slower output is obtained due to a process or a machine.

It might also be due to reworking a bad batch or batches, or repeating work which has failed.

All through this book, it is pointed out that any undue pressure coming from supply problems, impinges on, or could impinge on, quality.

One of the chief causes of errors in the Pharmaceutical Industry is caused by employing temporary staff who are not properly or adequately trained. Therefore, everything has to be done to reduce the use of temporary staff.

Temporary staff may however be necessary for economical reasons, to cope with events such as planned surges. In this case, all the precautions taken by proper training should obviate quality problems. The mere fact that it is a "planned act", takes any undue pressure out of the situation.

The author has seen on a packing line, 8 girls out of 9 as being temporary. This proportion, unless the people are really well trained and experienced, is not a healthy situation, and can lead to serious quality problems. The reason for this high number of temporary staff was that a tube supplier was 3 weeks late in delivery. This forced the packing department to make up for lost time in order to avoid an out-of-stock situation.

Critical Factors:

- Forecasting,
- Capacity – especially capacity for surges and unplanned production, not forgetting Infrastructure capacity, etc...
  - ease of putting in 2nd and 3rd shift if not already in place,
  - ease of working 7 days a week.
- Flexibility of staff: – terms of competence,
        – terms of availability,
- Management cover of all shifts,
- Training of staff: – generally,
        – in time for NPI
- Temporary staff fluctuations,
- Temporary staff training.
- Adequacy of equipment:
  - shortest possible change over times,
  - robustness
  - sizing
- Maintenance of equipment
- Speed of availability (negotiated with supplier) of equipment
- In Process Control
- Quality of incoming material: – specifications
           – control
           – storage
- Process robustness
- Eliminate reprocessing by finding the root cause of problems
- Factory and services reliability and maintenance,
- Factory and services maintenance scheduling and planning. Also its relation to long medium and short-term forecasting and planning.
- A culture of commitment, accountability, and reactivity.

Quality Actions:

This relates to all the quality actions and monitoring that is carried out in all factories by Quality Assurance systems and other means.

The supply chain puts pressure on virtually every part of a factory, its equipment, services, direct and indirect staff.

Any weakness, either technical or administrative will soon be revealed when the Supply Chain is first studied. More will come to light as the new methods of working, planning, and stocking are put in place.

Critical Quality Indicators:

- Recalls,
- Complaints,
- Lead times,
- Lead time variations

The whole Quality Assurance systems revolves around this – (see previous paragraph).

Of special reference is In Process Control and Quality Control – see next chapter.

Other indicators include:
- rework levels,
- temporary staff levels,
- temporary staff use frequency.
Client satisfaction indicators.

9 | *CONTROL OF QUALITY*

The Control of Quality intervenes at some, or all, of three internal stages in the supply chain:

- reception of goods or components in the case of the non certification of a supplier,
- during the process,
- prior to release.

Usually these stages are on the critical path. That is to say, the product cannot be further processed unless the testing is satisfactorily completed. Therefore, obviously there is pressure to make these interventions as rapid as possible.

It is therefore necessary to carry out as much of the testing as possible concurrently with other manufacturing activities. Therefore, In-Process Control (I. P. C.) often can and does substitute totally or partially for end and release control. On the other hand, certain tests cannot be completed quickly. For instance, sterility tests and environmental monitoring tests take a lot of time in order to obtain results from which decisions can be taken.

Quality control, either performed in masked (concurrent production) time, or as a result of immediate transfer of information, is a major contributor to agile manufacturing.

(Sterility tests are a good example where Parametric release (see later) often involves less risk and is more significant than actual sample testing).

Limitations of Control:

Quality cannot be controlled into a batch or a product. Quality is built into the product by the design, process, validation, people, method, procedures, etc...

Quality Control can never give absolute results. The proper validation of processes, both at the supplier and in-house, and sustained monitoring of the critical factors in manufacture, is just as vital if not more, then analytical control. These critical factors include people, their training, their motivation as well as documented and understandable procedures. Obviously, they include processes, equipment, cleanliness, and mechanical services such as air conditioning, vacuum, and compesed air and water systems, etc...

Double Control of outside purchased Materials

Any double control is totally useless as for instance is the case of a raw material being controlled both by the supplier and then by the manufacturer. Double control means time and money wasted.

Sufficient trust must exist between the supplier and the manufacturer that the assay results and other Quality Control results are accepted by the receiver. Proper audits and aleatory sampling should give sufficient confidence.

Some regulatory authorities do not accept this concept. They do not accept the sharing of responsibility, and they believe that the pharmaceutical firm should be held totally and uniquely responsible. This, provided that the supplier has established himself as reliable, is an outdated and unenlightened stance (see Chapter on Regulatory, page 159).

Similar arguments hold for shipping finished products from one firm to another, and from one country to another provided that no change or alteration takes place during transport, which again has to be validated. (see page 136, on Retesting).

Changes during transport could be induced by temperature changes and variations (see Chapter on Transport). Raw materials being shipped must, of course, be tamper proof.

The situation as regards shipping to emerging markets and certain countries must be watched with care. In this case, double checking and double control may be advisable.

Critical Factors:

- Good and accurate planning integrated with production planning,
- Robust and reliable analytical methods,
- Simultaneous information flow between Quality Control, In Process Control and Production,
- Clearly defined timing of tests (i.e. all links in the chain should know how long a test takes).
- When testing takes place outside of the manufacturing site (i.e. subcontracted or outsourced) clear sampling procedures and rapid transport arrangements are necessary.
- Agreed Certificate of Analysis tests.
- Certificate of Analysis – rapidity and completeness,
- Definition of accountability and responsibility,
- Validated Parametric Release procedures.

Quality Actions:

- Strict acceptation (Pass/Fail) procedures which cannot be overriden.
- Clear awareness by Quality Control staff of production cycle (time or value) and their own position in it.
- Awareness by other partners of possible bottle-necks in Quality Control.
- Trend analysis of repeat testing and reasons for these.
- Clear definition of acceptation/failure criteria (some tests are critical to product quality and may take a long time to obtain).
- Electronic Certificate of Analysis and signature.
- Parametric Release: what parameters.
- Training programmes.
- Training records.

It is sometimes much more significant and sure to pass or fail a batch based on information concerning a number of parameters associated with a batch, rather than actually testing the batch for a particular assay.

In the case of sterile products, the number of samples which would be required to test for sterility in order to get statistically significant figures, is so high that it is neither economical or practical to do so. Therefore, by taking a certain number of parameters around the filling process, sterile room, washing process, and sterilising process etc... one can see better the surrounding conditions in which the product was processed. This makes it easier to judge the likelihood of sterility.

By continually monitoring these conditions results are being obtained continuously, thus batch release can be obtained much faster. The actual sterility test results can be obtained only after 10 days of incubating in agar and other media to observe whether a germ grows.

The fact is that it is most important to prove that the process is continuously under control, and is in line with a validation protocol, that no unusual events take place and that if they do take place, the right corrective action has been taken. This is better than taking a sample and ascertaining that there is no contamination in it. It must be born in mind that the actual testing process itself induces a risk of contamination.

Proper validation, and proof of the following of validated methods, is a much better guarantee for product homogeneity and conformance to standards than actual sample testing.

Parametric Release for non sterile products

Certain other products such as tablets, creams, and ointments, could follow the same logic as explained in the previous paragraph. In process control (I.P.C.) data obtained throughout the manufacturing process is much more significant than taking samples and testing at the end of the batch. Corrective actions, recognising and dealing with anomalies, are much more important than bare results which are probably statistically not significant.

These facts are not as yet approved by many government inspectorates, mainly due to lack of experience, confidence, trust and technical knowledge.

Indeed, some authorities do not even accept I.P.C. results, they require other than production staff to sample and to test. This again shows a lack of trust and experience, and is often guided by outmoded corporation protection prejudices.

From the supply chain point of view, parametric release is an important help to reduce cycle times, and avoids the cost of double or sometimes triple testing.

The author is not advocating parametric release for all products all the time. A judicious choice with proper evaluation of risk is necessary for all products.

In-Process Control

In-Process Control (I.P.C.) is one of the most important methods of ensuring the good quality, or required quality, of products. In the case of a technical problem it enables rapid corrective action to be taken. Furthermore, it enables reduction of end testing, and therefore has a direct impact on reducing cycle time.

It does not matter who does the I.P.C., whether it be Quality Control or staff integrated into production, provided the responsibilities are clearly defined. From a supply chain point of view and a product flow point of view, it would be more efficient to have the I.P.C. activity integrated into production. This would enhance ownership and responsibility of the production staff, and also reduce lead times. Although, as mentioned above, unfortunately some inspectorates do not allow this.

Critical factors:

- Properly designed In-Process Control relevant to product quality,
- As much as possible automated recording of results with visible signals showing up anomalies,
- Disciplined recording of results,
- Ease of visibility of results, microtrends* and their communicability.
- Defined decision/responsibility structure.
- Trend analysis and its feed back to internal and external suppliers on specification control parameters.
- Validated Procedures.

Critical Quality Indicators:

- Length of tests (as planned)   – for incoming items
                                     for in-process
                                     for finished goods
- Real length of tests,
- Number of retests,
- Recalls,
- Resampling (a number of times),
- Deviations analysis,
- Batches rejected,
- Supplier's goods rejected (number of suppliers, number of goods),
- Length of shipments under bond,
- Time of arrival of Certificates of Analysis in relation to arrival of goods.

Batch Dossier – Supplier and Customer

Every batch of a pharmaceutical product must be properly documented, by which means one knows where all the ingredients

* microtrends find here really their most immediate effect.

and components have come from, what their analytical results were, as well as the I.P.C. results, the work sheet, and the final pass certificates of analysis.

One of the reasons for this is a need for "traceability" in case of a problem which could entail product recall (see page 137).

Shipping under Bond

One of the classical ways of saving time in the production cycle is to ship goods during the time that analytical or Quality Control tests are being performed. This is often done within a single company when a product is manufactured in one factory or country and shipped to another country or warehouse. For this to happen, it is essential that the Quality Assurance systems are in place and that they work satisfactorily.

It must be understood that under this principle there is a considerable pressure to release the product. This pressure can be diminished or reduced by clearly stating the time involved in the testing. The risks are well worth taking, provided that this information is known by all concerned, and that the system cannot be by-passed.

Naturally, total trust, transparency, and the sharing or defining of responsibility, is particularly important and useful in the case of a long analytical Quality Control time such as sterility testing. This is especially the case in the retesting of products presenting anomalies on the first test.

In order to optimise the supply chain, reduce errors, and simplify the administrative procedures, the use of electronic transmission both between supplier and manufacturer, as well as, between manufacturer and final customer, greatly facilitates the creation of the batch dossier.

Although in the past written signatures were necessary on most documents to show proof of veracity, most inspectorates now accept the use of electronic signatures.

Certificate of Analysis:

The Certificate of Analysis (C of A) is the bond within the supply chain which guarantees the quality of a product from one factory in the chain to the next.

Here two major problems arise:

First, the speed with which it is issued, despatched, and received. Very often, we find that the Certificate of Analysis arrives a week or two weeks later than the actual product, thus rendering it unusuable for that period of time.

The second is that the testing schedule and/or limits, and the way limits are defined, do not correspond to the registration dossier – the receiving plant generally being in another country. Therefore, further testing may have to be requested from the originating company's factory. The alternative being that supplementary tests will have to be performed by the receiving company.

It is true to say that if a factory makes a product to one specification, or to one set of testing criteria, for possibly five countries it naturally finds it difficult to have to supply further or different analytical data to an extra country. But until harmonisation of dossiers are achieved, this situation has to be coped with. To be able to provide extra information, the analytical department has to know which product is destined for which country, and/or if the product is shipped from a common store.

There is a lot of pressure in order to set up an electronic signature of computerized Certificate of Analysis, or even electronic sharing of analytical results. This kind of improvement, if carefully used would undoubtedly help to reinforce quality in the supply chain and also improve agility in lead time.

Retesting:

Retesting, in this context, refers to the retesting that occurs when a product has already been tested in one factory, a certificate of analysis established, and the receiving factory then undertakes further testing.

Generally there are two reasons for retesting.

One is a legal requirement where the final packer or "treater" of the product has the legal obligation to retest and assume final responsibility for release of the product. They must also provide proof of the quality of the product. This is the case in France where the paradox can exist where more than one factory can come under the same "responsible person" or Pharmacien Responsable and yet retesting is still required.

Often the problem is made worse by the fact that legislation requires sampling on arrival i.e. the samples cannot be taken and sent in advance for retesting. This not only means loss of time in retesting, but means time and space lost in resampling and also re-shelving or re-palletisation on arrival at the second point of control.

Only slow and methodical convincing of the authorities can obviate this point.

The responsibility aspect is absurd if this has to happen within the same company – owning plants in different countries.

It obviously demonstrates a lack of trust if retesting has to be done between two separate or associated companies. It also shows a lack of the sharing of responsibilities and in this case, both companies should pay for this inefficiency.

For example, the cost of the retesting of *one* product per year can amount to 10 x $ 5 000 for the actual test, plus about 10 x $ 500 for double materials handling, double storage space, as well as stocks.

This makes a total of $ 55 000 for one product.

If the company handles 15 products, the waste induced could possibly amount to $ 825 000 per year – off the bottom line.

Traceability

In the Pharmaceutical Industry, it is both a legal and a commercial necessity to have complete traceability of a product and its components.

As mentioned before, this is necessary in order to be able to recall a batch, or a number of batches, with which some problem could have occurred subsequent to dispatch from manufacturing.

Batch numbering systems and their proper recording all *along the chain* ensure this. The Quality Assurance system in the Manufacturing or Quality Control functions are the holders of this information.

In the case of the repacking or shipping of intermediate packs, it is also essential that the guarantee of traceability be maintained.

10 | *STOCK CONTROL*

This section refers to raw materials, work in progress, and finished goods.

The information diagram below Figure 10.1, shows the main interaction between the different parameters regarding stocks.

The important point is that information should constantly update the data bank. This database should have a built-in warning system which must trigger a rectifying action or, at least, a decisional process in the event of a breach of one of the parameters.

It is also important to realise that a mistake, incorrect information, or slow information, can upset the entire equilibrium within this diagram.

This, in turn, would put undue pressure on one or more parts of the chain, which could endanger both the quality of the product, and also affect the service.

Distributor and Client Warehouse Stock (V.M.I.)

Ideally supply chain implies that the manufacturer not only controls his own stock, but also that of his customer. By controlling, we mean knowing their stock and replenishing it at an agreed frequency, or within a maximum/minimum agreed level of stock. This is called Vendor Managed Inventory (V.M.I.).

If the "customer" belongs to the same company, then the information flow should pose no problems provided the computer system is compatible and linked.

The problem arises when the customer is either another company and/or a distributor who does not have a computer link, or the computers are unable to communicate with each other.

Yet this is one of the key elements in an agile manufacturing system and a successful supply chain!

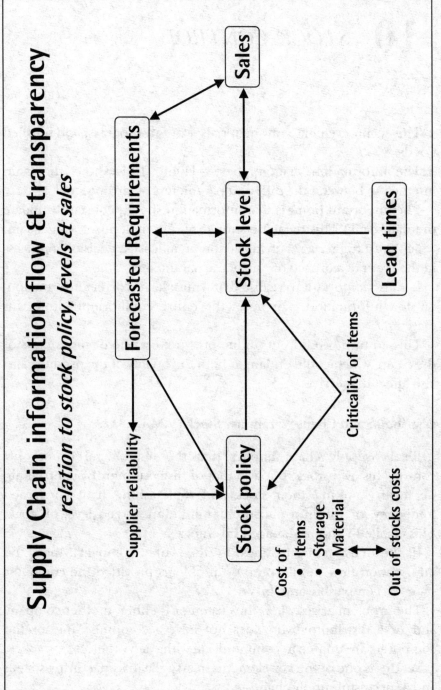

Supply Chain information flow & transparency

relation to stock policy, levels & sales

Figure 10.1

Persuasion, conviction, and investment, are the key methods to show that a common computer language is good for both businesses. On the other hand it is understandable that a wholesaler, who may have 150 suppliers, cannot have 150 different IT systems.

There are of course many other advantages in managing or controlling the clients warehouse; most notably the rapid access to commercial information leading to micromarketing.

Another advantage is immediate feedback to production and procurement – who can then act accordingly. This is particularly relevant when seasonal variations occur, or when sudden competitor action influences product sales.

It is generally in the client's interest to have low stocks. Since the manufacturer has the same interest, the risks of both running at low stock levels are great. These risks must be shared, as well as the responsibilities of possibly running out of stock or of carrying higher stocks.

Space within the customers warehouse is also a factor which has to be known in order to control his stock.

Intimate and sustained contact should be maintained in order to avoid unpleasant surprises, and to keep things running smoothly.

Stock Levels

Supply chain theory and J.I.T. theory says that stock levels should be at a minimum.

This is easily said, but what is minimum? Before answering this question, the parameters influencing stock should be enumerated:
- Actual sales,
- Forecast volumes,
- Cost of materials,
- Cost of storage,
- Space (availability and cost of outsourcing storage),
- Lead time and "Rarity",
- Cost of Quality Control,
- Risk of obsolescence and/or shelf life,
- Risk of non conformance on receipt and Fragility,
- Risk of non or late delivery (due to I.R. or other)

Figure 10.2, tries to capture this.

This diagram also shows the interactive and complex nature of the supply chain.

The subject of stock levels, minimum stock levels, and re-order levels, is so important that it is worthwhile examining some of these parameters.

Actual Sales and Forecast Volumes

These two parameters must be viewed together in order to assess variations and tendencies between the two. Obviously, in case of change, one has to react accordingly.

Cost of Materials

The absolute cost of materials is an important consideration as regards cash flow. 5 000 cartons costing $500 would have a different influence on stock values than would 500 000 cartons. Likewise one ton of raw material costing $3 500 would have a different effect on stock to 500 grams costing $300 000.

Cost of Storage

Cost of storage is related to interest having to be paid on bought and paid stock, and is also related to the actual cost of storing in terms of space utilisation – see next paragraph.

Cost of Storage Space

This can be an important parameter if there is shortage of space. If there is plenty of space available, such as is the case when a new warehouse has been recently built, then there is no premium on space.

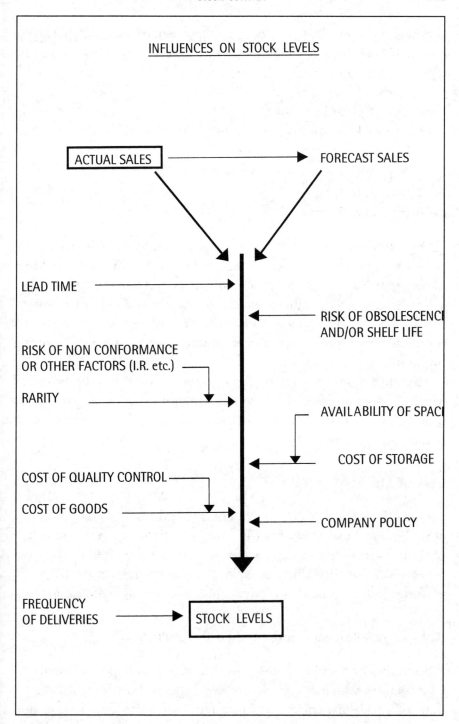

Figure 10.2

When there is a shortage of space, and where the available space has to be carefully alloted, or in the case of space having to be rented externally – then the cost of space has to be calculated in relation to:

  a) cost of outside renting,
  b) cost of keeping less stock and possibly thereby having to pay for more frequent deliveries,
  c) risks associated with keeping less stock.

Lead time and Rarity

One should always have more stock of products or items having long lead times than products whose lead times are short. This is necessary in case of production problems or sudden surges in sales.

Rarity refers to products which are either difficult to make or require special planning to produce. This is the case for seasonal products or imported products which have quota or rationing problems.

The author knows of a particular gold coloured aluminium foil which is only produced in limited quantities once a year. If an order is not placed early enough, one might have to wait 10 months before the next delivery !

Cost of Quality Control

Quality Control work on incoming goods, can represent a considerable burden both in human and financial resources. It is obviously less costly to analyse five tons of a raw material, than ten times 500 kg of the same material. The analytical tests are the same whatever the size of the delivery and likewise the sampling costs.

Risks of obsolescence and/or shelf life reduction

Some materials have a short shelf life. An example is Phenol or Chlorocresol which is used as an antiseptic and a preserving agent, both of which are cheap, and one tends to order larger quantities of them, in order to reduce analytical control costs. Yet the short shelf life mitigates against this. Printed material can become obsolescent due to marketing or regulatory changes.

Having large stocks which are never used and have to be subsequently destroyed – costs a lot of money, wastes space, and involves administrative, and subsequent destruction costs.

Risks due to non conformance on receipt and Fragility:

The stock levels need to take account of the risks of non conformance of the product at the time of receipt and/or at the time of use. Some deliveries might not meet specifications, or they might be spoiled during transport.

Fragility refers either to physical fragility where cartons or bottles are fragile and could be affected by adverse transport or weather conditions; or it might refer to chemical instability of raw materials or adjuvants. Examples of the latter could be colouring agents or flavours.

Obviously, errors of printing, colours of cartons and labels, can also present risks, especially from new or untried suppliers.

Risks of non or late deliveries due to other factors:

These risks refer to unexpected transport problems such as accidents, fires, or more importantly, industrial actions such as strikes or go-slows.

These have to be evaluated as a function of the environment, and possibly the social climate both at the supplier and the receiving firm.

It is fairly evident from the above that Stock Levels have to be set or computed:
a) item by item,
b) supplier by supplier,
c) taking into account the environmental situation,
d) financial consideration,

No simple or single magic formula exists.

Out-of-stock Situations:

As said above, the prime objective of the supply chain is to have the right product at the right place, at the right time, and at the right cost. The biggest failure, or ultimate breakdown, of the supply chain is an out-of-stock situation. (Indeed this, with quality issues, is the main cause of failure of the manufacturing function).

There can be a number of reasons for out-of-stock situations:
1) Inadequate capacity anywhere along the supply chain.
2) Breakdown of equipment or plant along the supply chain.
3) Forecasting outside agreed and shared figures.
4) Error in planning and/or distribution.
5) Breakdown of planning, either inappropriate or inadequate Information Technology capability.
6) Stock levels too low due to bad management, financial constraints.

These matters have been dealt with in the last few pages and can be considered as acute out-of-stock situations.

As these problems are rather aleatory, unforeseen, and rare, the only answer is to have back up suppliers or outsourcing capabilities, or a rapid and agile reaction.

There are however sometimes chronic out-of-stock situations.

Chronic out-of-stocks:

These are much more serious in nature and as they could concern a number of different products and packs and might occur frequently. Here the problems is one of mind sets and attitudes. In this context, there are two kinds of mind sets. (A mind set is a conditioned state which can eventually become a dogma).

1) Budgetary Oriented Mind Set,
2) Value   »   »

In Budgetary Oriented Mind Sets (B.O.M.S.), the technical director and *his staff* are operating with their eyes fixed on the budget both in terms of investment as well as operating cost. Their time is spent in cost cutting, analysing, adjusting, debating for hours a 10 000 $ saving, etc... Generally overtime and extra cost incurrements involve heavy and long administrative and authorising arrangements.

This mind set is obviously not suited to rapid increases *or* decreases in demand, nor agile investment and capacity increases.

In Value Oriented Mind Sets (V.O.M.S.), any decision on investment, overtime, and other operating costs, are viewed with profit loss or gain in view. "If we don't do this or don't deliver that, how much value shall we lose?" "If by investing this, even at a premium price, how will it affect our bottom line"?

In this mind set, the sharing of cost and opportunities with marketing or other parts of the organisation is a current method of operation. Just as the sharing of the risks of forecasting is done with marketing, so the sharing of investment could also be considered.

This value oriented mind set concentrates on adding value, and on the overall aims of the business, rather than a functional or local interest approach, where compartmentalised issues take up the sight of vision. The sort of questions that are put by this kind of mind set are: How can we work over week-ends? How can we make our machines more productive? Would output increase if we put an extra machine or person on the line? How can we reduce losses, even at the expense of putting in more resources into the line? Could we transport this by air and still make a profit? Time is spent on creativity and inventiveness instead of counting cents.

B.O.M.B. and V.O.M.S. are two different cultures. The agility of the technical staff shows itself by moving easily and quickly from one mode to the other! However both require to be practised at different times and under different circumstances.

Generally, in a chronic out-of-stock situation, it is the Budget Oriented Mind Set which operates. It is a cultural feature, it operates throughout the technical organisation and the supply chain. Drastic and radical measures are to be effected to turn the mind set into a profit oriented state.

Chronic out-of-stock situation can last for years unless top management changes are rapidly brought about.

BUDGET AND PROFIT MIND SETS

	BUDGETARY ORIENTED MIND SET	VALUE ORIENTED MIND SET
– INVESTMENT	– According to budget.	– According to pay back and production.
– OPERATING COST	– "	– In relation to Value Added and work load.
– STAFF NUMBERS	– Never exceed budget.	– In relation to work load. Use of temporary staff, Short and medium term. Flexible, multiskilled staff.
– AGILITY	– Weak.	– Strong.
– ANTICIPATION	– Weak.	– Strong.
– REACTIVITY	– Slow.	– Fast.
– ATTITUDE	– Rigid.	– Flexible.
– WORKING HOURS	– Strict adherence.	– Flexible, modulated.
– STOCK LEVELS	– Predetermined.	– Variable.
– OUTSOURCING	– Yes, if budgeted.	– Yes.
– VISION	– Department Oriented, sectorial.	– Company Oriented, holistic.
– TIME SPEND	– Counting Costs.	– Seeing where resources are best deployed.
– RISK	– Little risk admitted.	– Shared and evaluated.
– TRUST	– "Trust the budget".	– "Trust the sales"

Figure 10.3

Figure 10.3 compares some of the implications of the two mind sets.

Of course, it is up to C.E.O.'s to inculcate the right balance between the budget and the value mind sets. This does not only apply to the technical function and the supply chain, but also to marketing and any other departments which should rapidly react to opportunities or better still to anticipate. In fact, budgetary mind sets sterilise anticipation and the thinking ahead of wider possibilities than those dictating on-going strategy.

Other out-of-stock comments:

There is a third reason for a chronic out-of-stock situation and this is also a type of mind set, but is related to historical events. If a company has underinvested in the past on a new product and subsequently the product sold well, an out-of-stock situation arose. It is likely that for the next new product, a more liberal approach to capacity investment along the entire supply chain would follow. If this product sells well, then there is no problem. If, however, it does not sell well, and thus causes an under-utilised overcapacity, then all is not well. Reproaches, regrets, and blame fly all over the place. It is highly likely then, that at the next new product launch, capacity investment will again be restrained. Hence an out-of-stock situation arises, if the product sells well.

This is a sort of high-low cycle which permeates the company mind set.

The common element running throughout these mind sets is "FEAR". Fear of overinvesting, fear of overstocking, fear of exceeding budget, fear of not fulfilling shareholders expectations, etc...

Recklessness in building too much stock and over-investment can of course lead to disaster. Finding the balance between the insufficient and the excessive is the art of management and the use of experience. But a fear culture can never produce lasting and sustained success.

To close this section on out-of-stocks, one other point should be mentioned. The production management and the supply chain management should be as near as close as possible to its market, in order to get the "feel" of it, and to be able to take its "pulse". Distant and often centralised decision-making does not favour good stock control.

Out of stock situations due to unexpected regulatory constraints, sometimes induced by protectionist attitudes, are not considered in this guide.

The Supply Chain Reaction (Inverse Forrester Effect)

In a situation where buffer stocks do not exist, the slightest hitch with a link in the chain can affect the entire chain in a whiplash (see page 336) or wavelike manner where the negative effects are accentuated and thus increase exponentially along the chain.

Let us take an example. A supplier delivers faulty cartons, the strip packed tablets to be packaged are ready, and so is the cartoning machine and packing line. The consequence is that the tablets cannot be packed, the packing line lies idle, the staff on the packing line have no work, and the tablets in question become out of stock.

This is the immediate effect, but of course the chain reaction starts. When the good cartons come in, another product was scheduled on the packing line.

The choice then has to be made as to which to pack. Overtime or an extra shift, probably at extra cost, makes up for the incident but in the meantime, the tablet press has also had to stop, as no stock build up is permitted. This then creates the same backlog effect on the tablet press, tablet room cleaning process, laundry, etc...

So we can see from what is a simple carton defect the following supply chain problems are created:
- scheduling,
- capacity wasted,
- product out of stock,
- other product or products out of stock,
- staff waste,
- increased costs.

Critical factors:

- Knowledge of Lead Times,
- Calculation of optimal stock levels,
- Accurate knowledge of stocks at all locations:
  - supplier,
  - in house
  - at subcontractor
  - warehouse
  - client
  - in terms of:
    - volume
    - value
- Rapid knowledge of stock movements,
- Policy
- Procedure/system for modification of policy
- Materials handling equipment and installations
- Strategic stocks
- Definition of critical physical conditions of holding stock,
- Space availability of stock – capacity (especially for seasonal or other fluctuation),
- Definition of where and how much stock is kept,
- Bottle necks at goods in and out,
- Transport and weather conditions,
- Maximum variation of sales,
- Risk factors.

Quality Actions:

- Definition of critical stock items,
- Reduction and simplification of stock movements,
- Auditing of supplier stock control
- Help with client stock control
- Flexible stock policy modification
- Proper control of physical stock movements

Critical Quality Indicators:

- Stock figures,
- Stock movements,
- Supplier status,
- Capacity (physical) status,
- Monitoring of physical/critical conditions,
- Lorry and goods movement fluctuations,
- Stock adjustments,
- Inventory accuracies,
- Out of stock records,
- Retest records,
- Supplier reliability records,
- Stock turn round

An holistic understanding of the factors mentioned in this chapter, and those mentioned in Chapter 2 is necessary to reduce lead times and have optimum stock levels. Only thus can one pinpoint to levers for improvement and agility. Focussing on a single or few factors only, could result in a negative overall effect on the business.

11 DELIVERY

Delivery obviously is on the critical path of the supply chain, both for goods coming in and goods going out. Proper, safe, cost effective, and reliable delivery systems and processes must exist. However it must not be thought that a good supply chain means necessarily Just-in-Time delivery for every product every time – see later section on Transport.

The right delivery conditions depend on the optimization of cost and time factors, also the integrating of these with other links in the supply chain.

Choice and cost of transport

Delivery times can vary from 3 to 5 times between one transporter and another (all using the same method of transport). By using different methods of transport, variation can vary from 10 to 30 times depending on what shipping method is used. No wonder then, that J I T delivery is, to a very large degree, influenced by transport and their arrangements.

The cost of the goods, plus their transport costs, can also vary to a large extent. So in order to optimise total cost benefits, it is necessary to compare the goods and transport costs with the service levels required, such as precision and timing of delivery.

Air Transport

While transport by air is generally quicker (especially for long distances) the cost can be prohibitive, especially for low cost products. In some countries, storage in airports is not safe, thefts might occur, whereas a lorry driver rarely leaves his cabin.

Pressure and temperature variations in aircraft holds can have dramatic effects on some pharmaceutical products: pressurised Aerosols are a case in point. Guarantees must be obtained either from the transport company concerning conditions, and/or trial deliveries must be undertaken. One must bear in mind not only the total time, but also cycling time i.e. how many times the aircraft stops off, and where, on the overall journey. Further stability testing might be necessary to ensure product quality.

Road Transport

Even over short distances of 300 to 3 000 km, road transport is longer, but cheaper. It is important to be aware that this length of time could create problems when passing through cold temperature zones. Overcooling could have just as much harmful effect on a product as overheating. These zones could affect the stability of some pharmaceutical products.

Similar barriers can be caused by mountain ranges. For instance, the Alps or the Pyrenees in Europe; or the Urals between Europe and Asia. These barriers not only create uncertainties of time, but they can at different times be cold zones or very hot zones – which can also affect the products being transported.

Bad roads, bad road conditions, or poor road networks, can also retard delivery times. Crossing the English Channel also slows down transport times.

The cleanliness of lorries is also an important quality criteria. Sometimes, many financially attractive offers are obtained in cooled or temperature controlled lorries. On further inquiry however, one might find that the pharmaceutical goods would be transported with raw meat. If the lorry is not meticulously cleaned after each delivery, smells and bacteria could accumulate. Smells can affect even fully packed pharmaceuticals, and bacteria do not make good travelling companions.

Customs:

Crossing borders between countries can have a retarding effect on transport due to customs and/or police controls. Often, these delays are unpredictable. For instance, a lorry from France to Turkey could take anything from three or seven days.

Under these conditions, it is therefore difficult to establish policies in terms of strict stock or, for that matter, delivery time, quantities, and similarly lead times. Policies and lead times must be such as to buffer the effects of transport condition variations.

It is important that paperwork should be filled in *meticulously*, otherwise delays can be counted not in days but in weeks.

The clients' input on custom rules and regulations are vital if problems are to be avoided.

Size of Shipment:

Size of shipment also has an affect on direct shipment costs. These must be weighed against waiting for optimal shipment size. This could engender some items in a shipment to be late – thus missing a customer or a sale. This entails extra stock cost. Optimization of shipping is not necessarily optimising the supply chain and reducing overall cost. Per unit, transport costs must always be traded off against total logistic costs and in short, total costs, which is another example of the Value Flow concept. (see p. XXIII)

Standardisation of shipping:

The standardisation of shipping between supplier and manufacturer, and between manufacturer and customer is a key supply issue.

Shipping not only refers to the method of transport chosen, but also the way the product is multiple packed, grouped, palleted, labelled, etc... The customer in this case may be the manufacturer's own warehouse, a sister company, or an outside distributor.

Standardisation generally concerns:
- codes, bar coding,
- labels,
- outer cartons,
- pallet sizes,
- pallet loads,
- pallet heights,

It often also concerns the total number of pallets shipped, and sometimes the batch sizes.

The speed of reading codes and checking items is increased severalfold if labels are in the same position on cartons, and if the code is the same throughout the chain. Codes and quantities can be fed automatically into stock. The danger of transcript errors is reduced, if not totally eliminated, and this therefore is a very important quality factor.

Labels which are not coded can present tremendous potential dangers. Even colour differentiation does not give full protection against errors. Errors can occur at both the dispatch and receiving end of the manufacturing cycle.

Outer cartons should be the same size. This facilitates handling at dispatch. Furthermore, it allows pallets to be stacked in a standard manner. Speed and space are gained, and quality is improved.

The author has seen many pallets poorly stacked because there were two different size outer cartons of the same product (coming from two different factories in the same group).

Instability can lead to many accidents, where not only human injury can be sustained, but product can be lost. A carton of 2 dozen bottles of syrup does not need to fall from more than one metre before it breaks! A breakage of this nature can ruin thousands of other packs of other products.

The supply chain would thus be very adversely affected, not to speak of damage to the quality of the products. Repacking and retesting takes time, increases costs, and jeopardizes quality. Although this sort of incident is insured, the low stock environment prevalent nowadays might require the recovery of as much as possible of the undamaged goods, hence the repacking operation.

Whilst standardisation should be as widespread as possible, good supply chain management requires cost optimisation and this in turn may require non standardisation. As an example, let us take the case of a standard pallet having a standard number of boxes being shipped (by air) to Russia.

During the summer, large aeroplanes are used for the journey, and these are able to take pallets loaded up to a height of 1.75 metre.

In the winter, however, smaller planes are used, requiring either smaller pallets or pallets not higher than 80 centimetres.

In this latter case, the packing lines and the instructions must be adapted to the seasonal air traffic requirements! An example which also shows the complexity of the supply chain, and the flexibility required to manage it.

Critical Factors:

- Reliability in timing
- Temperature/physical control if relevant during transport
- Choice of transport
- Customs
- Transport regulations
- Definition of delivery times: – period
                                 – hour
- Cost of transport
- Material handling
- Adequacy of transport
- Knowledge of climatic conditions during delivery
- Knowledge of climatic conditions at arrival (geography)
- Industrial relation problems en route
- Political problems en route

Quality Actions:

- Trial deliveries
- Product testing after transport

Critical Quality Indicators:

- Delivery reliability records
- Late deliveries
- Complaints from customers

12 | *REGULATORY*

Regulatory Requirements touch virtually every part of the supply chain.

Most regulatory requirements purport to have quality as a reason for existing. As the pharmaceutical industry is totally controlled by regulatory authorities, it can be said that every aspect is critical to quality.

The industry is regulated in most countries by the following ministries:
  – Ministry of Health,
  – Ministry of Environment,
  – Ministry of Social Security,
  – Ministry of Works,
  – Ministry of Finance,
  – Ministry of Transport,
  – Ministry of Air.

It would be too long, and perhaps unnecessary, to enumerate every aspect of all these ministries which have an impact on the supply chain. The fact that there is a speed limit on every road in most countries, which reduces the speed of transport and thus could lengthen the delivery time, is a minor case in point.

But certainly transport laws impinge on delivery times. In many countries, lorries are not allowed to circulate during the weekends. In some countries, social laws do not allow factories to work or drivers to unload goods at weekends. Whilst these factors impinge on the supply chain, they do not necessarily have an impact on quality unless the resultant additional strain put on the supply chain subsequently causes breaches in quality. An example would be a consignment which arrives slightly late at the end of the week, and because of the need to unload the lorry before the weekend break, the driver helps to discharge the goods. As he is not experienced, a pallet is dropped and goods damaged....

The main regulatory implications, however, stem from the Ministry of Health (the Food & Drug Administration (F.D.A.) in the U.S.A.). A pharmaceutical product cannot be put on the market without a product licence. Obtaining a product licence takes a very long time and is a complex procedure in every country.

Apart from the safety and efficacy aspects, which are thoroughly examined by the authorities, there are some of the following subjects which have to be examined and approved. They all have an impact on quality and on the supply chain.

- Formula,
- Specifications of ingredients,
- Specifications of packing materials,
- Process,
- Source of ingredients,
- Source of packing materials (some),
- Analytical methods,
- Analytical norms (standards),
- Identification of tablets,
- Printed text on labels,
- Printed text on cartons,
- Printed text on leaflets,
- Information technology systems.

Most authorities have to approve the place of manufacture, including the stores and finished goods warehousing of the manufacturer (or whoever holds the product licence). Many authorities inspect these locations on a regular basis. Among other things associated with the supply chain, they look at:

- Product Flow: • Physical,
                         • Administrative.

- Layout
- Physical Conditions of   · Cleanliness,
  Manufacture, Packing   (absence of vermin and beetles)
  & Storage   · Hygiene,
                             · Tidiness,

- Returns,
- Supplier Audits,
- Transport Methods,
- Batch Reconciliation,
- Production Losses,
- Certificate of Analysis,
- Batch release authorisation,
- Batch release documentation.

They do not at present as yet look at:

- Lead Times,   – Abnormal Demand Procedure,
- Cycle Times,   – Abnormal Supplier Problems,
- Output,   – Abnormal Demand Process,
- Costs,   – Comparative Purchasing Costs,
- Efficiency,   – Batch Sizes,
- Scheduling (push or pull) – Capacities

Both the product approval methodology and inspection methodology are generally described (often indirectly) in codes of practice known as Good Manufacturing Practices or G.M.P.

Whilst a factory in one country is supplying that same country, the situation is fairly straightforward. When, however, a factory is supplying different countries the problem becomes complex. The regulations and G.M.P. guidelines often differ, not in every way but in certain respects. Some of these differences can be significant and, indeed could, be contradictory.

Let us look at a few examples, which are especially related to the supply chain.

1) In most countries the authorities allow a product to be accepted for use (passed) by a company or factory, if it has been tested in another factory belonging to the same

company. This is not the case in Belgium and France where the product has to be retested. This not only costs extra money, but also consumes time.

A batch of raw materials worth $ 1 million may have to wait a week for analytical approval – the cost of waiting runs into thousands of dollars. If we multiply this by the number of raw materials a factory deals with and add to it the number of batches a factory receives, costs can rise to $ 2-5 million per year. This is not only extra cost and time but it is *wasted* and *wasteful* time as the product has already been accepted by the same company!

2) An excellent example is shown by the Certificate of Analysis which some countries accept and others will not.
   (see Chapter 9 on Control of Quality).

3) Some countries require the actual sampling of incoming raw and packing materials to be effected by Quality Control staff. Other countries require the samplers to be adequately qualified, but they may belong to the stores or warehouse department. A similar situation arises with work in process,or finished goods, where in most factories production staff – properly trained, take the samples.
   It is obviously cheaper, simpler, and faster, for staff working near the point of sampling to take the samples. The supply chain is lengthened by the extra time and often extra planning needed for the Quality Control samplers to be at the right place at the right moment. This procedure can be taken to its ridiculous extreme when Quality Control staff have to go into a sterile room to take samples. Not only the time, the care, and the effort necessary in washing and gowning up is wasted, but the product is put at risk by the extra bioburden of additional people going into a sterile area.

4) Some countries accept shipping under bond (see Chapter 9 on Control of Quality), others do not.

5) The analytical specifications of aspartame are different in

the U.S. Pharmacopoeia from some of the European pharmacopoeias.

For instance, one specifies a conductivity measure, whilst the other does not; one requires an ultra violet measure and the other does not. Also the methods of finding impurities is different.

This means that two complete series of analytical tests have to be performed, in order to conform to the two pharmacopoeial requirements. This consumes time, money, and other resources, and is essentially wasteful.

HARMONIZATION

In order to simplify and shorten the supply chain, the harmonizing of as many of the factors mentioned on page   is necessary. As mentioned in chapter 6, one cannot underestimate the importance of standardisation, and the earlier in the product's life cycle that this is established, the better. Harmonization increases the agility of the manufacturing operations to a tremendous degree. There are two distinct categories of a product's characteristics which should be harmonized.

There are aspects, such as the formulation of the dose form or the specifications of the excipients, which should be fixed during the design phase of a product. These should be standard for all markets and not be changed, because this will require additional clinical trials and a re-registration process which is both time consuming and expensive.

There are also items which are defined at a much later stage such as artwork on the carton which is almost always not standardised between markets. These are items which, although under regulatory control, are more easily changed. However it does not detract from the fact that the more one can harmonize a product's physical or presentational characteristics – the simpler the supply chain becomes.

Let us look at each in turn.

Harmonization of formulae

It is often difficult to understand why the formulation for the same product should be different. Yet it often is so.

There are two main reasons. The first is that in some countries, one of the ingredients, or some of the excipients, is not allowed. For instance, this is true in the case of Bismuth which is permitted in some European Union countries, but not in others.

Another reason is that some countries or markets do not like a certain colour or taste. Yet another reason is that some countries prefer a different texture for their creams or ointments. These reasons are generally historical and refer to older products. Recently developed formulations are generally more standard.

The difference in formulation is very minor, and of course the efficacy of the product is *not* affected. Yet, if a certain formula is registered and accepted by the authorities, it is difficult to change. Good reasons have to be given, evidence of stability has to be provided, and even new, lengthy, and expensive clinical trials may be necessitated.

A typical example is the case of a cortisone ointment where one country's marketing required the presence of a very small amount of lanoline. This is not allowed in most other countries.

Another example is peppermint flavouring, which is very much liked in Anglo-Saxon countries; whereas orange or strawberry flavouring is preferred in latin countries.

Overages, for reasons of stability, which are permitted in certain countries, but not allowed in others.

As far as the supply chain is concerned, a different formula is as different as another product. Its manufacture and packing has to be planned for separately, and equipment has to be cleaned between batches just in the same way as if it were another product.

If one wishes to improve the supply chain, the harmonization of different formulae for the same product are, therefore, of capital importance.

Simplification of formulation also reduces the chance of errors of product mix-up, and thereby quality is more easily obtained.

Harmonization of Specifications for Ingredients

Specifications for active ingredients are generally the same throughout the world; although some excipients have different specifications.

An example is sugar, where in some countries the kind of impurities allowed is not the same as in other countries.

Another aspect of specifications is limits. Generally, tolerances are given to quantities of active ingredients and excipients found in the final product. These limits may be plus or minus 10 % or 5 %. These limits are allowed due to manufacturing variability, losses during manufacture, and indeed due to the precision of the analytical methods used in finding the right quantities. Finding precisely and accurately 50 micrograms of a corticoid in a single aerosol dose is no small analytical feat.

Products having different ingredient specifications, and different limits in product specifications, require separate handling in the supply chain.

In fact, we could find that we have four different "products". It often means having two nearly identical excipients serving exactly the same pharmaceutical purpose. It could mean having two different kinds of finished stock for exactly the same use, both being equally efficacious.

Tight specifications are sometimes registered in order to avoid generic houses copying the product. This does not always succeed. When the generic house registers his product, authorities may well accept a looser specification, knowing full well that the effectiveness and usefulness to the patient is the same. Another reason for giving tighter specifications to the authorities may be ignorance on the part of the Registration Department of a company who might argue that internal limits (which are generally tighter) can always be achieved – so why not register it.

Another problem with specification is sometimes either the lack of it, or the imprecision of it. An example is lactose. Lactose is used as a diluant and excipient in tablets and dry powder formulations. Its properties allow powders and granules to flow more or less easily in hoppers and silos. Yet the crystalline characteristic of the lactose is not specified either in pharmacopeia or registration documents. In fact, it is difficult to characterise by physico-chemical means. Experience is the best way. Many manufacturers have made batches of tablets or powders and, from time to time, found it a very difficult and slow process on account of differences in the physical properties of the lactose. These are recurring problems which slow down the production cycle – hence they impinge on the supply chain and, of course, on the quality of the product.

Another example, which in this case has a double action on the supply chain, is sodium citrate. This is also used in some tablet formulations as an excipient. It is a product which has to be used within a short and limited time after its manufacture, otherwise its crystalline characteristics change (crystals grow) causing its flow properties to deteriorate.

It is of course possible to change the specifications at, or from the supplier, but this is not always easy. The pharmaceutical industry uses very specific and special products. However it does not use them in large quantities when compared with other industries like the food or soft drinks industry. Citric acid is a case in point. The soft drink industry uses this in hundreds of tons per year, whereas the pharmaceutical industry may require only half a ton a year. Thus it is hardly worthwhile for the producer to change his specification for such small quantities.

Harmonization of Specifications of Packing Materials

Packing materials, notably those in contact with the product itself, have to have registered specifications. Examples include glass bottles, vials, ampoules, rubber plugs, aluminium/plastic foil, plastic and aluminium tubes for tablets, ointments, etc...

Again different regulatory authorities may have different views on the use of packing materials. The tolerance of lead in glass, which are expressed in parts per million of lead, differ from one country to another. Certain additives in plastic containers used for tablets are not permitted in some countries.

These factors increase the difficulty of managing the supply chain where harmonized packing material presents obvious advantages.

Another example is child proof packaging. What is accepted in some countries as safe from inadvertant use or opening by children, is not accepted by other countries. Plastic foil for strip packs containing tablets or granules is a case in point.

Harmonization of Manufacturing Processes

Processes are often developed with the intention of using a particular kind of equipment or set of equipment which is then registered with the authorities.

Different manufacturing sites may have different equipment and hence the processes registered may be different. In the case of rationalisation or multiple source from strategic sites, the supply chain may be seriously affected if re-registration of different processes is necessary.

Re-registration can take a long time and, in some cases, it is not allowed. Very minor changes sometimes create problems. For example the length of drying registered for a granule – may be 45 minutes. With experience it was found that after 30 minutes, the product was sufficiently dry, and also gave better compression properties. Nonetheless, this was considered as a process change and re-registration, with stability data, had to be made.

The whole philosophy of process improvement, part of the continuous improvement programme in most factories, rests on shortening cycle times whilst improving quality. This is often jeopardised by heavy re-registration procedures.

Continuous improvement principles are entrenched in other progressive industries such as computer chip manufacturing.

Another example is the case of using high shear mixing for tablet masses as against using fluid bed mixing. Analytical results and stability results may be identical, but as one method was put into the registration file, the other cannot be used prior to re-registration.

Harmonisation of the Source of Ingredients

Most regulatory authorities require not only the registration of the specifications, standards, and analytical methods for an ingredient, but they also require to know the source. This is notably true of active ingredients which, in many cases, are patented and manufactured in special purpose primary plants.

Registration of the source of primary material frequently means, in depth inspection for these plants and regular re-inspections. Any proposed change of sourcing has to be notified to the Authorities and they have the right to re-inspect the new source before authorising the use of the product. The process and method of manufacture, the size of batch (if made in batches), still has to be seen prior to approval, even though the standards are identical with the first source.

When we talk of other raw materials which might be manufactured by a number of suppliers, the authorities must be informed of the source or sources used in a formulation. The authorities may inspect these manufacturers.

It must be realised that legally the pharmaceutical manufacturer of the *final* product is responsible for the quality, purity, and source of its input material.

The major regulatory impact on sourcing is that no switching or substitution can be done – unless it is registered. Registration of a second source can take many months, and this must be taken into consideration in the cycle time of the production. This is especially true in the case of capacity constraints. There are some countries which will accept only one registered source at a time, thus adding inflexibility to supply.

The Certificate of Analysis (see page 85) emanating from some countries is accepted by some, but not others. For instance the Certificate of Analysis from within the European Union is accepted by other EU countries, yet it is not accepted in countries outside of the EU.

Because many pharmaceutical Groups have their primary plants outside the EU, eg in Puerto Rico or Singapore, their Certificates of Analysis are not accepted in Europe. If the company has two sources, one within EU and one outside and sources from both, then two different stock items have to be carried for these very same regulatory reasons. Again, unless both sources have been declared in registering a product in a certain country – the two products cannot be used interchangeably.

Harmonisation of Analytical Methods

Analytical methods have to be registered at the moment of seeking product registration. If, after experience, one wishes to change – then the method has to be registered again.

A product may be registered in one country in one year, and in another country two years later. If in the meantime the method has improved, obviously the newer method is registered. So one might well be in a situation where a factory produces for the two countries, but having to analyse the same product using two different methods. Thus increasing the cost and also reducing the flexibility of using the same batch for the two countries.

An example of the loss on the drying of ranitidine effervescent tablets is tested differently in Europe from the US. This obviously creates additional tests if one wishes to have a common stock for both clients.

Analytical Standards

Standards registered in one country might also differ from those registered in another country. This is again generally due to the timing of the registration. Therefore separate stocks of the same product might have to be carried.

An example of this is the weight content of a certain tablet which was originally registered for one country at ± 5 %. Two years later the registration department of the company registered the same product for another country at ± 3 % because, by experience, that was the internal limit used and could be attained. They in fact tightened the official limit. Later the product was made on another machine where these higher limits could not always be obtained. The result was that every third or fourth batch was unusable. Months had to elapse before the standard was re-registered. It is always much more difficult to justify any widening of limits rather than the contrary!

Printed text on tablets

Tablets are often identified by codes or letters and/or figures. This may be done by embossing or printing the tablets. Some countries neither need nor want these identifications. One can easily imagine the loss of time in changing over from one set of punches to another, as one has to switch from making from one country to another, the changing of punches on a tablet press could take 3 hours. The advantages of harmonising the embossing or printing is therefore fairly obvious. Generally one can persuade these clients who do not want this kind of identification on economic grounds.

HARMONISATION OF LEAFLETS, CARTONS, AND LABELS

Harmonisation of these items from one country to another (if the language is the same) would present considerable product flow, quality, cost, and supply chain advantages. Multilanguage leaflets and cartons are already much in use, although still not acceptable everywhere. In some countries a foreign text, even if the native text is present, is considered illegal. There are fortunately many areas which can be harmonized, such as carton size, but proactive action is needed and, of course, any changes have to be registered.

The time it takes to approve a multilanguage leaflet or label can be very long, this obviously impinges on the supply chain and can be critical in a launch situation where speed can be very important

Harmonisation of printed text on labels

Bottles, ampoules, vials, even tubes, and sometimes strip packs are labelled. Labels are either pre-printed or printed on line on the packing machine.

The texts give formulae, instruction for use, name and address of the manufacturer, as well as expiry dates and batch number.

Identification is generally a legally approved text. It is approved for each country. Every word, figure, comma, and even layout has to be submitted to the authorities. Once approved, no change (not even a comma) can take place without a renewed approval.

These facts apply not only to the final pack to be sold, but also to sample packs. These, as far as the end of the supply chain is concerned, are different products.

Size of the label is generally governed by the size and nature of the object on which it is to be attached. Marketing has also an influence on size and text disposition.

Obviously any change required by the authorities or by the company (such as a change in the name after a merger) requires new authorisation and takes time to obtain.

Generally government authorities give time for the stock in hand to be sold or used. However if large stocks exist, repacking is necessary if it is worthwhile or has to be done – otherwise the product has to be discarded.

Harmonisation of printed cartons

Very much the same principles apply to cartons as to labels. However changes in artwork, due to different size or change in colours, is more often requested either for commercial or marketing reasons. Sometimes a change in the pack is required due to new equipment or new size of packs initiated by production.

Stock levels, and quantative requirements over a given period of time must be calculated in order not to cause waste.

The efficiency and speed of the artwork department is vital in the supply chain. Approval of texts within the same company can take many weeks. New systems of transmission of artwork by fax have facilitated this task. Indeed, nowadays, it can be done on a computer screen interactively between two countries.

Generally alot of pressure is put on the artwork department to work fast. However this raises a very critical quality point where lots of errors can, and do, occur. This is a key link in the supply chain when it affects new products and/or a change in pack.

A factory supplying 100 different packs to 25 different countries has virtually every day a new pack or pack change to deal with. Because these are often in unfamiliar languages – they make the task more difficult and challenging.

Harmonization of printed leaflets:

Text on leaflets are very strictly regulated by the authorities. No unjustified claims to effectiveness must be stated, and they must conform strictly to the indications for which the product licence has been obtained.

Leaflets are either precut or they come in rolls and are cut just prior to being inserted into cartons. The incidence on the supply chain of the time for approval are similar to those applying to cartons.

Harmonization of Information Technology Systems:

Regulatory authorities look at all IT systems related to production. Not only must validation data be available, but inspectors ask specific tasks to be demonstrated in front of them in order to show the security within the system.

See chapter 13.

SUPAC: Scale-up and post approval changes

Certainly in the U.S., the SUPAC Guidelines are having a major impact on filing post-approval changes. These Guidelines are facilitating site transfers and impacting the way in which New Drug Applications are being written. The SUPAC Guidelines are setting standards/expectations for classifying equipment, determining the impact of making changes to product formulations, adjusting batch size, etc...

Critical Factors:

- Specifications,
- Overspecifications,
- Supplier approval,
- Supplier procedures,
- Storage conditions,
- Delivery conditions,
- Manufacturing Quality Assurance systems,
- Environmental protection,
- Social and Works law conformance,
- Transport constraints,
- Working hours constraints,
- Traceability guarantees,
- Complaints

Quality Actions:

- Audits
- Supplier training
- Traceability systems
- Procedures
- Environmental monitoring
- Calculate chain speed as a function of regulatory constraints

Critical Quality Indicators:

- Supplier audit reports
- Complaints
- Inspection – Internal and External Reports
- Regulatory instructions (complaints)

13 | INFORMATION TECHNOLOGY AND AGILE INFORMATION FLOW

Information Technology is not only the key to the supply chain – it is the pillar on which it rests. It is the only way to have visibility of material flow, inventories, and demand. For it to be successful, it must read demand instantaneously at every level, and most importantly at the final level. This, in the pharmaceutical industry, is the pharmacy and the hospital.

Information Technology is the only means by which the early warning tracking system can exist (see below).

The major positive aspects of the impact of Information Technology on quality related to the supply chain are as follows:
- Errors of manual transcription are reduced if data is transferred electronically.
- Reproducibility and consistency of data by the use of validated programmes and algorithms are obtained.
- By the speed of data flow, every link in the chain has access to information quickly. This obviously improves the quality of business decision making.
- The sharing of the database (i.e. everyone using the same information) avoids misunderstanding.
- The traceability of goods is assured, because the transactions are based on the same system.
- Client/server design allows data to be viewed interactively in different geographical locations.

However a certain number of points require careful consideration:
Any central Information Technology system must be maintained in alignment with local systems. This is because misalignment can give

rise to incorrect information. An example is Bills of Material for conversion of sales forecast into component supply.

Design constraints should not have an impact on the flexibility to respond to changing business needs. For instance, if the system specifies transport lead time by customer, rather than individual item, then improving customer service could be constrained. A large quantity of data might not necessarily mean a lot of information.

There must be a manual back up system in case the Information Technology system breaks down. Recovery systems must also be in place.

The validation of the system must be end to end. Local interfaces and translation tables must be reliable, because one weak link in the chain can affect the whole.

False security may be engendered by blind reliance on any Information Technology system. Trained, skilled, and competent staff are still required. Staff who can question the data both to identify errors, and to recognise real business changes.

Internal and external customers and suppliers should be involved in the establishing of design codes and standards. The big problem might be, and generally is, that a supplier cannot afford multiple customer systems.

Finally, it must not be forgotten that Regulatory Authorities might wish to inspect, test, and have proof of the validation of an Information Technology system. This is particularly the case where Certificates of Analysis and batch information are stored in the system. These issues are of particular importance in the "paperless factory".

Early Warning Tracking System (E.W.T.S.):

This system exists in the head of every good planner. It does not exist as a computer system. Some people say that the algorithm necessary to cope with the complexity of the information would be too difficult to compute or it would be too expensive.

If we refer to Figure 2.1 of the Supply Chain for an Aerosol, we can see that for one finished pack of aerosol, there are about 50 processes, about the same number of components, and possibly 50 suppliers and sub suppliers involved. Any one of these suppliers and processes can run into problems, without mentioning transport, customs, import licences, and other logistical problems.

It would therefore be normal that a signal flashes all along the chain if there is a problem at any one of these points.

When we consider that a finished pack may have many clients, i.e. different language packs or a different number of items grouped together, the complexity of the information network can be imagined.

The system must operate both from suppliers to manufacturers, and from manufacturer to client and, of course, in both directions.

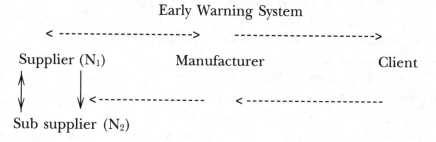

Early Warning System

This is a system which allows maximum agility in the manufacturing operation and, of course, the whole extended supply chain.

The following are considered to be the main constraints in establishing an Early Warning System:

– The need to recognise significant deviation from monthly sales profile,

– The need to recognise and differentiate the product or product family,

– For the data base to be truly significant there must be a good statistical and historical base,

– The need to recognise seasonal forecast (this can be fairly easily planned),

– The need to recognise changes which are not caused by either change in demand or problem of supply (such as exceptional purchasing terms or strategic stock build up, etc...).

– The need to recognise the input of a new supplier having the same stock item (code),

– The need to be set to the correct sensitivity:

– if too sensitive – many spurious warning messages would occur

– if too insensitive – will miss changes.

The Information Technology system would have to be carefully aided by people who are able to recognise or analyse market intelligence.

This system should not be substituted for an "Abnormal Demand Process" which is a management process that recognises abnormal supplier, logistic, technical, and demand changes; and which can evaluate the impact of the changes on the supply chain.

With Vendor Managed Inventory an Information Technology system warning planners of change must exist. This is part of the Early Warning System.

A firm's ability to react, either to abnormal demand or abnormal supplier problem, is an excellent indicator of its capability of managing and controlling its supply chain.

Information Technology can and must help this. Nevertheless, it cannot be a substitute for people input and managerial decision.

AGILE INFORMATION FLOW

Agile information flow as mentioned previously is very important in the supply chain, the whole concept of immediate market feed back, that this book refers to, is part of it. But the flow does not only apply to marketing information, it applies to production, delivery, supplier, and customer information, etc... The network for each, must be well traced, practised and sometimes tested to prove its speed and reliability.

An agile information flow has the following characteristics:

– As soon as the information is generated it is transmitted to those

interested, or at least to those who the emitter thinks is interested, at the fastest possible speed (measured in minutes!).
- If the emitter does not know who is interested he transmits it to someone who does know, who then in turn transmits it.
- It must be transmitted irrespective of hierarchical levels.
- It must be transmitted irrespective of functional divisions.
- It must be in an understandable language.
- Transmission is not enough, assurance of receipt and having taken cognisance of the information is necessary.
- It must have feed-back, so that the information is understood.
- It is concise, clear, accurate, and as complete as possible.

The e-mail fax, letter as such is not fully agile, the receiver might not look at his e-mail, the secretary might put the fax or the letter in a pile and the recipient may not read it for days. The only sure system is the telephone, where direct contact with one person is obtained, this is not agile, however either, as other people who should be receiving the information simultaneously do not automatically receive it.

Not all information is urgent, staff should be trained to discern the urgent, the important, and the less important.

It is difficult to train a computer to discern between:
- urgency,
- those interested in particular,
- importance.

An algorithm could be made, but agility means unpredictability, unexpectedness, and speed. By the time the programme has been made, the problem or subject cannot be treated in an agile manner.

Agile information flow is a key competitive tool in all aspects of running a firm, even an efficient I.T. system is not substitute for it, yet it obviously enhances it.

Agile information flow is again a cultural matter. Managers in an agile organisation have this automatic reflex and know when, and how, and to whom, to pass on what information, and make sure that the information is received, understood, and acted upon.

In order to retrieve stored information the system must be user friendly.

Critical Factors:

- Security of information,
- Integrity of information,
- Capability to interconnect with other systems,
- Manual back-up systems,
- Recovery systems.

Quality Actions:

- Full utilisation of data,
- Understanding of information.
- Reliable historical base,
- Item code records,
- Item code update systems,
- Abnormal demand systems,
- Test procedures,
- Agile flow.

Critical Quality Indicators:

- Warning of information loss,
- Validation data,
- Traceability tests,
- Information flow – degree of agility.

14 RISKS, RESPONSIBILITIES, TRUST, AND LOYALTY

RISKS

Three main types of risks are in question:

a) Supply breakdown

b) Quality breakdown

c) Information sharing and confidentiality of throughput figures.

In order to minimise any risk situation, it is best to share it. One can only share something if one knows what it is, and therefore the assessment of risks by both partners in the chain are necessary. Sharing must start with the mutual communication of corporate goals of *both* parties concerned.

As we go down the chain, the consequences of the risks increase in value. Therefore each linkage must have its risk value which means different types of contracts for suppliers along the chain. Similarly, as volumes or turnover change, so the risks also change. Therefore they must be reassessed.

The potential risks of supply problems and quality problems are being treated in different parts of this Guide.

Confidentiality of information is a far more tricky subject. Few contracts, if any, are completely safe and watertight. As quantitative information is generally available on computers, many people can have access to these. Thus the pinpointing of responsibility becomes virtually impossible.

An industry wide transparency scheme such as E.C.R. (Efficient Customer Response) is one answer to this problem, but the pharmaceutical industry is far too closed and secretive to partake of or to create it.

Risks have to be evaluated bearing in mind the unexpected and the unpredictable (see pages 12 et 28).

Anticipation which is an integral part of agile management has also to be assessed in terms of risks. The cost of response weighed against the cost of non response is a matter of shrewd judgment and must be shared by different functions. The maximum experience and knowledge of the supply chain and its impact must be exercised.

Critical Factors:

- weigh pros and cons of a single source of supply.
- pros and cons of a supplier knowing trade figures.
- degree of confidentiality required.
- controlability of confidentiality.
- minimising risks by knowing mutual constraints:
    - production:
        - (reliability and age of equipment)
        - process
    - delivery
    - financial
    - storage
- insurance contracts.

Quality Actions:

- Definition of responsibilities
- Definition of sharing   – consequences
                     – losses
- Regular review of risks in showing the evolution of volumes and sales.

Critical Quality Indicators:

- Turnover
- Financial state of supplier
- Delivery track record

RESPONSIBILITIES

Legal as well as moral and commercial responsibilities are involved in the link between the supplier and the manufacturer; likewise between the manufacturer and the distributor.

The success and effectiveness of the supply chain depends on the commitment of the staff of suppliers and distributors. Their thorough understanding of the role and responsibilities is essential. They must understand not only the needs of their client, but also the needs of the final client – which in this industry generally concerns an ill patient.

Critical Factors:

- Definition of legal responsibilities,
- Definition of commercial responsibilities,
- Making suppliers understand clients needs all along the chain,
- Traceability systems.

Quality Actions:

- Training of supplier staff,
- Quality programs at supplier,

Critical Quality Indicators:

- Complaint records,
- Delivery track records,
- Internal quality reports of suppliers.

The subject of responsibility in the pharmaceutical industry is very large. It is not the subject of this book. The large nature of the responsibility is fully understandable, because lives are at stake. This makes the job of all those working in the industry and those associated with it, the more challenging and demanding.

TRUST

Trust is more than likely to be the main management theme in the next decade.

Competition and trust are two notions which appear contradictory and incompatible. This is more so in business and trade where, in the past, one could only succeed to the detriment of the "other". Today, companies realise that they not only succeed together, but they realise that by working together they can attain mutual benefits.

If we look at history, the same pattern emerges.

In prehistory, man taught man to gain possession of property, chattel and land. Later, they banded together to form clans – because they found it more effective to reach their aims.

Then clan fought clan, to gain territory, power and goods. Many of these clans became kingdoms. Then kingdoms fought for wealth and influence. Subsequently empires collided in global conflicts, sometimes in gigantic dimensions and with horrendous results.

Now, most nations realise that by working with one another, synergies, and mutual benefits can be found.

In parallel with these martial events, three phenomena have developed over the last two hundred years.

Firstly, man's awareness of his culture, and the analytical examination or study of man's development as a social and economic phenomenon.

Secondly, the rise of capitalism, the creation of large, and then mega firms, some of them on a global scale. And in parallel, perhaps partly in consequence, the rise of communism or anticapitalism. This has given rise to a new logic.

Thirdly, the comprehension of the human mind, of human nature, and the psychology of behaviour, has been refined.

These three threads have not yet, to this author's knowledge, been brought together in a simple, clear, analytical form.

Not only has it not been brought together and viewed in this three dimensional manner, but it has not been examined in the modern business and management context (A notable exception is the excellent book on Trust by Fukiyama, which is wide reaching but somewhat heavy and perhaps too detailed to be practical).

Trust then has at least three main components:
- Logical (rational),
- Cultural,
- Psychological.

as shown in Figure 14.1

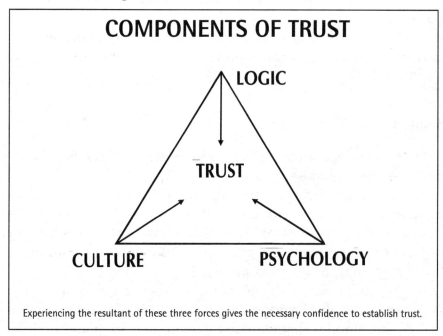

Figure 14.1

We shall try to analyse these three in turn, especially with a view to business and management.

Logical

There are two opposing logics in trust.
The one is that trusting someone else is not worthwhile as the

other's interest are either against our own, or it is too risky, in that the other might break "his word or contract" of that with which we have trusted him, as he has no interest is keeping trust.

The other is that once an agreement has been reached between two (or more) partners, it is logical that each should be held to it, and hence each can trust the other.

The way most Japanese firms operate is based on trust, not only between firms and their employees but between firms. This is a very strong component of Japanese's business success.

In Japan, trust is both a logic and a culture, it is ingrained in all dealings and is a matter of honour equivalent to loyalty and patriotism.

In western eyes, this factual and successful experience of trust may give a reasoned (and therefore logical) proof that trust is worthwhile!

Cultural

In every society or country, there are different ways that trust is perceived and practised.

There are those who are trusting; and those who are suspicious depending on their cultural and historical background. Of course, all the gradations in between exist. Most people are somewhere in between and thus act in different ways at different times according to the circumstances.

It is not the place to discuss here which people or which races or nationalities display what degree of trust. All that is required to be understood is that there exists a cultural component, which might be strong, weak, or variable, depending on the type of people we are dealing with and which may differ according to different circumstances.

Only experience can influence or modify this cultural component.

But by accepting these facts, one can weigh the impact of trusting or not trusting, or to the degree to which one can trust. Then in consequence, act or react in a conscious manner.

Psychological

Behaviour science has shown that animals can be taught to trust – they can be "tamed" and taught to perform different exercises involving a partner. The partner may be a human being, who may let him down and thereby create hardship or ill. Man knew this and has done this for millenia, but he has understood the process less well than now.

Experts say that trust is derived from the experiences of a child during its first year of life. This suggests that it depends on the childs maternal relationship.

Trust as a Mind Set

Trust, as many basic psychological concepts, is a question of mind set or attitude.

By this we mean that the mind is tuned to a certain set of thoughts, or has a pattern of thinking, say that one is climbing a hill and thinks that one should, or must, get tired, then automatically sooner rather than later, one does get tired – irrespective whether this tiredness is real or not.

One finds what one expects! it is another way of describing a mind set.

If one anticipates trust, one is much more likely to find it!

Distrust

If one does not trust, and one doesn't expect the other to trust, then we are in a vicious circle of distrust.

Unless we can persuade first of all ourselves to break this circle, we cannot expect the other person to trust, or act, in other than a distrustful manner.

In other words: trust engenders trust and, distrust engenders distrust.

Building up trust

Confidence gradually rises by building up trust in a gradual and progressive manner based on honesty, integrity, and transparency. What is ultimately required is confidence on both or all sides i.e. mutual confidence. (see Emotional Bank Account on next page).

Once the experience of mutual confidence has been established and shown to be sustained, the psychological block either ceases or is considerably attenuated.

The Meaning of Trust

Trust means faith in someone or some thing.

Etymologically the word Trust has it roots in Truth.

Trust in French is "Confiance" which comes from the word Confidence = with faith

Therefore Confidence and Trust have nearly interchangeable meanings.

In fact, it is RELIABILITY which is mutual UNDERSTANDING PLUS AGREEMENT.

Trust is sustained and fulfilled expectations. It may therefore be partial or progressive, and does not necessarily have to be total or complete.

Trust and sharing of Values

Trust can only be obtained by sharing; sharing of values, or of culture, of religion, of family ties, business network interests, or whatever. Sharing means expecting something *automatically* in return. The key word here is "automatically", or in other words implicitly.

Covey* says that trust is the highest form of human motivation. In fact, trust is the key to delegation and empowerment.

Now the reason that often trust is broken, or not obtained, is that

---

* S.R. Covey, The 7 HABITS of HIGHLY EFFECTIVE PEOPLE
   publ. Simon & Schuster.

the partner does not fully understand what the other wants in return. Thus the "understanding" has either to be decreed, which is generally done by setting out things in a contract, or it is implied by agreement.

Again, Covey compares trust to an "Emotional Bank Account" in which a reserve can be built up by proving the use of trust. This reserve can sustain some mistakes thus giving some flexibility in relationships.

Trust in the Supply Chain

Trust can be a very strong component in supply chain management. Often it can determine the success of a chain. Whilst the objectives of partners might not be identical in all respects, it is essential to find the coinciding objectives, which should be the basis of negotiation prior to establishing trust. It is one of the most important and difficult aspects of the chain to put in place and subsequently to maintain.

It is important because a whole set of partners, within and outside firms, are concerned by it. Figure 14.3, attempts to capture this concept. If one fails – all might be affected. It is difficult because experience, both in life and business, has taught us to be more suspicious than to be trusting.

The whole concept to win/win negotiations is based on establishing trust.

But once established it must also be continued, because if one or two distrustful acts are made the trust is lost. Once lost, recovery is a long process.

The main problem is that it is very difficult to explain both why it is necessary to trust and also how to learn to trust. This is especially true in a centralised "low-trust" society where the level of confidence of people towards other people, companies towards other companies, is historically and culturally low.

For trust to be effective, there must be clear understanding of each others:
– interest,
– competence,

- integrity,
- financial situation,
- objectives,
- way of operating,
- visibility and availability of information.

There must be no hidden agendas.

For all of this, there must be a certain level of high sophistication morally, politically, and financially.

In other words, *BUSINESS ETHICS.*

There must be mutual self confidence where threats and weaknesses are known and therefore exposed to exploitation. Guarantees to counter these must be given. In Japan, these guarantees (if necessary) are often given by banks. The comparative size of organisations must be taken into account. A large company when dealing with a smaller one, must not use its leverage unilaterally.

Trust is a multidimensional necessity. The fact is that where there is dependance and interdependance of:
- work,
- responsibility,
- service,
- quality,
- financial, personal and business relationships,
there is a need for trust.

Although our theme is the supply chain, the same problem of trust exists:
  a) within a company between staff and management,
  b) between companies,
  c) between sister companies within the same corporation,
  d) between companies and the environment – government, ministries, localities, etc...
  e) between management and unions.

In order to carry out work optimally, there should exist a large degree of trust between all partners. (See Figure 14.2, Trust Partners in the Supply Chain.)

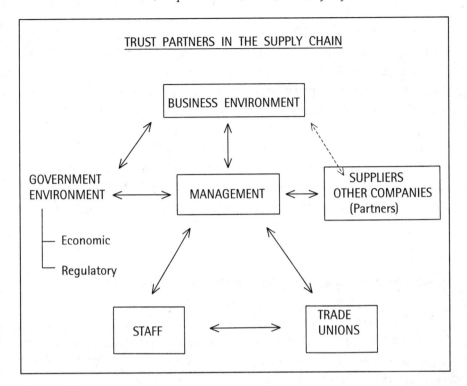

Figure 14.2

Trust as a tax

Distrust or mistrust in any business is a form of tax – thus creating an extra and unnecessary cost.

Trust and Loyalty

Finally, it must be added that Trust is the main reason for Loyalty, which is vital in any partnership.

The effect of trust on the quality aspects of the Supply Chain is at all levels

At manufacturing level, matters of quality and reliability are intimately related to trust.

The fact that a packing girl can be trusted to say that she has made a mistake; and that she in turn can trust her management not to sanction her – shows trust. The fact that a machine has been set so that it will produce more and the product will be ready earlier, and that the engineer can be relied on to inform the client is – a matter of trust.

The purchasing manager, the supplier, must be trusted as much as the goods receiving person and the mechanic who fits up a packing machine. The person who passes a batch of ointment must be trusted as much as the planner who is in contact with a customer.

Trust and Agile business techniques

The importance of trust in agile management techniques has been pointed out in Chapter 1. Trust not only reduces transaction costs but it also gives the confidence to act rapidly and take risks if and where necessary.

The need to spell out everything in detail, confirm and reconfirm with half a dozen signatures from different functional entities, is avoided in a trust culture.

Vertical Integration and Trust (Transaction Costs and Trust)

The whole issue of autonomy and complete "control" is all about trust.

"Trust yourself and your own sphere of influence rather than others", that is the reason or wish to control and have mastery over one's activity. Vertical integration is generally based on so called economic factors yet the real underlying factors, which are difficult to quantify and awkward to admit are due to trust factors.

These factors are in transactions and transaction costs. A lot of modern thinking about management and reengineering is about reducing transaction cost.

Aids and Barriers to Trust

Below is a non exhaustive list of potential barriers and aids to trust.

Potential Barriers to Trust	*Potential Aids to Trust*
Experience (poor previous)	Experience
Reputation (based on facts)	Reputation
Opinion (of others)	Credibility
Dislike	Like
Misunderstanding	Understanding
Lack of availability of information	Loyalty
Lack of Intelligence (or limited)	Sound argument
Opposite Sex	Transparency
Complexity of Business	
Language	
Ethnic:　　　race	
culture	
Religion	
Untidiness	
Body Odour	
Hygiene	
Physique	
Mental Frame (make up)	
Class	
Inflexibility	Flexibility
Incompetence	Competence
Uncertainty of market place	
Mistrust leads to mistrust	
Political beliefs	
Dogma	

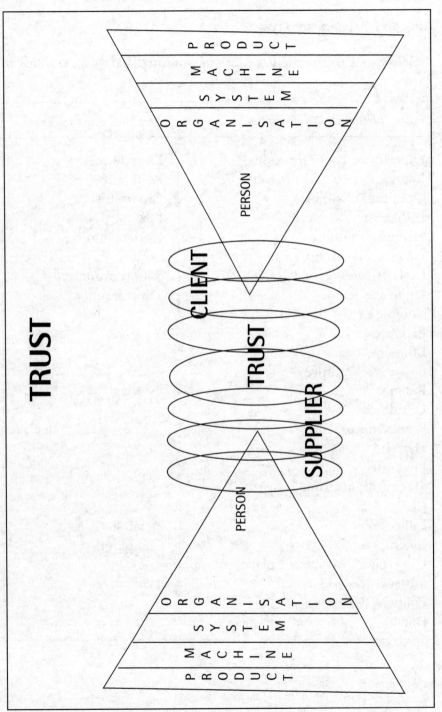

Figure 14.3

LOYALTY

Loyalty is one of the most sought after qualities which any person, company, corporation, institution, party, or country longs to possess and keep.

Yet it is an ephemeral, intangible quality, often expressed in unmeasurable ways. It is, like trust, a psychological phenomenon – yet it mostly has a logical and historical background.

For a pharmaceutical company and its supply chain, the players or partners involved, are:
– shareholders,
– staff,
– suppliers,
– distributors,
– pharmacists, hospitals, nurses,
– doctors,
     and above all,
– patients, or customers.
     (see Figure 14.4. The Loyalty – Trust Cycle)

We shall try to define why loyalty is important, how to create and maintain it, what its value might be, and how to measure it,

Like any other business factor, loyalty must be managed. If it is not managed, then it cannot be created nor maintained.

Far more significantly, loyalty can be split or divided in conflicting directions. This will occur between some of the players mentioned above. In this case prioritisation, and risk evaluation, are the managerial tools which need to be put in to play.

Customer Retention and Shareholder Loyalty

Most of the work on loyalty has been done in the context of customer retention i.e. what percentage of customers one retains over time. Whilst this is a figure which is relatively easy to calculate, it is nonetheless, a negative approach to loyalty. One counts those which go away, obviously an important fact to know, but it is more important to study the facts which:

a) make people come to you,
b) make people stay with you,
c) and most importantly make people stay with you despite some adversity.

It is in adversity that loyalty is vital, incalculable, and the sterling test for one's products, services, and the company's management.

Even the best managed companies have their ups and downs. Indeed, a company which continuously increases in size, profitability, and reputation, is probably mythical or non existent. Therefore, it is necessary in these "downs" to maintain customer loyalty, partner loyalty, and shareholder loyalty.

If one does not have this loyalty, then the company is forced to manage with a short term view by maximising profits for the short term. This does not create sustained shareholder or stakeholder value, which in this author's view is the most important mission of any organisation.

The supply chain which leads to efficient delivery is very much in an exposed situation. It is one of the easiest ways to create loyalty by excellent and reliable service to customers, and similarly the easiest way to break or weaken loyalty by poor service or poor quality.

Staff Loyalty

Staff loyalty is a two way process. It means management's loyalty to staff – and the inverse: staff's loyalty to the management and the company. One cannot expect the one without the other.

In any constrained situation, whether it be pressure due to work load, flexibility, speed, and the reactivity required for a particular situation; the normal carrot and whip is not enough to sustain the performance of the organisation. Loyalty gives the extra energy and will to overcome these constraints.

In a restructuring, reorganising, or re-engineering phase, loyalty is both the most precious asset and the most difficult to manage.

How can one be loyal towards the staff and expect the staff to reciprocate, if one has to lay-off a certain number of people?

The highest of managerial skills are required in such cases.

The subject is very similar to Trust (q.v.).

Empowerment does not lead automatically and immediately to loyalty. Autonomy, and being responsible for all of one's actions, leads to a certain self sufficiency and independance which does not require loyalty. However, over long term, empowerment (if properly applied) should induce loyalty because there is no total autonomy or total independance, – thus the loyalty effect is a powerful renforcement in any relationship.

Supplier Loyalty

As mentioned previously, trust and loyalty have been two of the key elements of the Japanese success story.

Loyalty between a manufacturer and his supliers is the final bond which makes for fair deals, smooth transactions, and full value to both partners.

Once established, it must not be abused. Short term gains by switching suppliers, especially without prior and timely warning, could lead to major quality and service problems in the supply chain.

Dominance, exclusivity, arrogance, and self satisfaction, are the chief enemies of loyalty and these attributes could be displayed by both sides.

Loyalty must be earned by negotiation, trust, and the mutual alignment of objectives (see chapter on Supplier relations).

Measuring Loyalty

Some of the metrics used in endeavouring to measure loyalty are listed below:

- customer retention
- supplier retention
- evolution of cost of purchases
- shareholders (number)
- value of shares
- value of company
- staff turnover
- staff satisfaction
- complaints
- customer satisfaction

It must be remembered that loyalty flows or fluctuates, and therefore these measures are not always relevant.

One should not hesitate to change the measures by inventing new and more appropriate ones.

Creating Loyalty

Loyalty is created over time, by reputation, by example, and by results, both financial and other appropriate figures.

Creating loyalty with staff is most readily obtained by empowerment, empathy, and transparency i.e. visible equity in emoluments, promotions, and status.

Genuiness and sincerity are also important ingredients to induce loyalty.

Creating loyalty with customers is by openness, fair dealing, good service, and rapid and complete information.

With suppliers, the creation of loyalty is again by transparency, sharing objectives and vision.

The common thread running through all of these is honesty and communication, both very similar to the business ethics mentioned in the section under Trust.

To sum up: Loyalty is a choice and a commitment, but it is temporal. Therefore, in order to sustain it, one has to manage and nurture it.

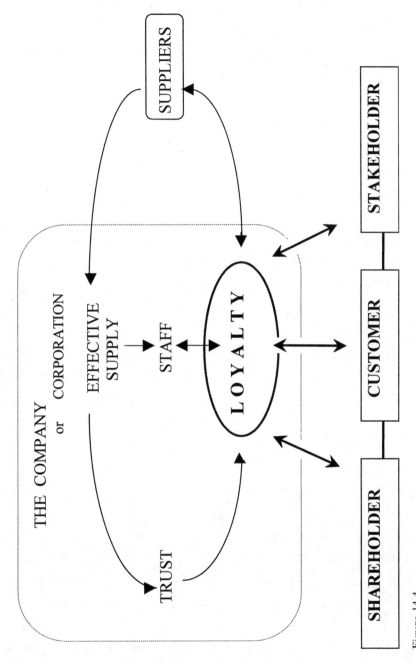

THE LOYALTY - TRUST CYCLE

SUPPLIERS

THE COMPANY or CORPORATION

EFFECTIVE SUPPLY

STAFF

LOYALTY

TRUST

STAKEHOLDER

CUSTOMER

SHAREHOLDER

Figure 14.4

15 | COMMERCIAL AND ORGANIZATIONAL ASPECTS OF THE SUPPLY CHAIN

Commercial

The commercial impact on the supply chain, and the consequences from it, is a large subject.

It includes:

- Toll manufacture,
- Transfer prices,
- Parallel imports,
- Currency variations,
- Quotas,
- Import and export licences,
- Customs and duties,
- Taxes,
- Market place information,
- Micro marketing,
- Reference pricing
- etc,...

However, few have any direct impact on the Quality of a pharmaceutical product. Nevertheless some will be mentioned.

Critical Factors:

- *Quotas, Import and Export licences might have time limiting constraints. This could put undue pressure on the chain.*
- *Customs could hold up deliveries, thus creating similar unplanned time pressures.*
- *Currency variations could alter critical choices in priorities.*

Parallel Trade

Parallel trade refers to buying a product at a low price from one country, then importing it into another country where the same product is sold for a much higher price.

The trader pockets the difference. Parallel imports and exports are the same thing, but looked at from two different countries.

The company selling the product is losing this difference in income.

Similarly, the loss of profit to the manufacturer is quite important.

One is generally talking about expensive products, otherwise it would not be interesting for the parallel trader to engage in the operation. A product which is selling for £ 30 a pack in the U.K. and for £ 18 equivalent in Greece, for instance, gives a differential of £ 12 per pack, of which the importer pockets perhaps £ 8.

At 100 000 packs, he makes a profit of £ 800 000 per month and this roughly is the sum that the manufacturing group loses. If we multiply this by a few products and a few countries we can soon realise that important sums are involved.

This sort of situation is fairly common in the European Union where there is a free flow of merchandise, but where prices are sometimes quite disparate.

Although regulatory authorities do not like this way of distributing medicines, they can do nothing about it. Unless, and this is the key to the issue, the product licence, i.e. formulation, etc... is not the same in both countries concerned.

A wholesaler, who may be the parallel trader, might order partly for his own country's use and partly for parallel exportations. He is not bound to inform the manufacturer of where the product's final destination will be, even if asked.

From the supply chain point of view, the system is fairly perverse, because essentially the manufacturer does not know where his product has gone. He has no right to refuse sales, even if he suspects that the product is going to be parallel exported.

It is against all the principles of transparency upon which the supply chain is built, not to know where and in what quantities one's production is.

More importantly however, there is often a risk to the patient. Dosage and other instructions could be in another language, thus endangering the patients' understanding. The patient could take the wrong quantities and/or at the wrong times and continue for a wrong period.

The colour and taste of the product may be different, thus inducing confusion. If, as sometimes is the case, the product is repacked by the importer, then the manufacturer looses all control of the integrity of the manufacturing cycle. Repacking and relabelling errors have occurred in the past. This is the main point where parallel imports could engender major quality problems.

Only tighter regulatory measures could reduce this problem, such as stricter repacking laws and stricter control of importers. Common pricing policies would totally obviate this problem.

Transfer Prices

The price of raw materials and finished products are established according to different commercial agreements between different countries.

Where this factor may have a negative impact on the supply chain is in a case where a manufacturing unit is making a product for two different countries, each of which has a different price for the same active ingredient.

Auditors, for accounting purposes, may require the same product to be stored and processed separately. Therefore the same ingredient, with exactly the same specifications, has two codes or reference numbers – this in itself can create confusion. If, as is often the case, a batch for one country is then followed by a batch or part of a batch for another country, in the same manufacturing vessel or line, then for efficiency reasons it is better to work in a continuous mode.

However, the accountants may wish for clear separation. This is a technical and quality nonsense.

To illustrate the point, one can imagine a person placing a $ 1 000 in a bank, and another person placing $ 800. Each would insist that the bank keep these moneys separately!

Accountants may argue that if more losses are incurred for one country's product than for another, it is difficult to calculate the value for each if the product or ingredient is mixed.

These are artificial arguments, leading to artificial problems which slow the supply chain and incur extra costs.

Reasonable auditors can, and should, be persuaded to accept the mixing of ingredients which have only nominal cost or price differences.

Toll Manufacture

Toll manufacture means manufacturing for someone else at a rate (toll) without paying for the active ingredient – which is generally expensive.

If the manufacturer makes the same product for his market, and therefore has to pay for the ingredients, the accounting becomes difficult, unless the ingredients are held and treated separately. In this case, a similar problem to the one mentioned in Transfer Prices arises.

Not withstanding these challenges, many companies or manufacturing units work with this method. It somewhat encumbers the supply chain, but is a viable method which may give rise to overall financial benefits.

ORGANIZATIONAL

The internal structure of a pharmaceutical company is as many and as varied as the industry itself, though broadly speaking they fall into two main categories. There are those companies which are centralised, where a central head office issues strong directives to a relatively small number of manufacturing units. Also there are those that operate on a decentralised management structure where power and decision making is, to a large extent, left to the local operating companies. It should not affect a well run supply chain if a company belongs to either one or the other of these categories. But there are certain factors to be aware of:

As insisted upon on numerous occasions in this book, the closer one is to the market, the easier it is to identify with its needs. The nearer staff are to their customers, the more they can anticipate and get involved with their requirements. Commitment, extra effort and that "feel" for the customer is much more easily obtained if there is a physical proximity between customer and supplier.

Agility in supply, manufacture and human relationships are also made easier by local or at least regional and frequent contacts between partners.

Closeness of customers and a decentralised organisation does, or should in no sense, interfere with global policies, global vision and global supply chain network and efficiencies.

On the contrary, local strengths should feedback and reinforce global efficiencies.

Decentralisation does not mean isolation! It is a cultural feature of a firm and is really a mind set.

CENTRALISATION

In this case, the supply chain should be managed, directed or piloted, by one central authority within a group. Responsibilities must be clearly delegated and defined. Also responsibility must cohabit with authority. Too often, the supply chain is not effective because authority and responsibility is diffuse.

If it is well defined and given full authority, it is not a contradiction to say that a team can pilot the supply chain effectively. In the case of poor structure, the impact on all aspects of Quality can be disastrous. It is particularly important to consider structures in a globalised context.

GLOBALISATION AND REGIONALISATION

The critical Quality aspects of Globalisation and to a lesser extent Regionalisation are that:

- Products might be shipped as intermediates,
- Local or Regional packing and/or labelling and/or relabelling takes place.
- Transport uncertainties can become more acute.
- Custom clearance and holding can create storage condition problems.
- Precise knowledge where products are and have been sent -traceability.

This all adds additional complexity to the supply chain, especially with many factories each supplying a broad range of markets with different products or dose forms both in finished packs and in bulk for local packaging, (it is well known that over 80 % of errors in the Pharmaceutical Industry are in the packing and labelling operation) therefore critical control must be exercised over this activity.

Furthermore, shipping unlabelled material, (unless coded under the label – implying local code reading) – is also a dangerous and "unpharmaceutical" practice. Retesting or recontrol practices have again to be critically controlled.

Economies of scale, however may render the supply more efficient and rapid.

Changes in consequence of Supply Chain Implementation:

Whether we talk of local, regional, or global structures, functional barriers have to be broken down when implementing the supply chain.

Entrenched systems of logistics, information technology, and procurement must be changed. Furthermore, while some systems or arrangements might have been efficient and cost effective – radical changes have to take place in order to unify the system.

Any change can induce quality problems – new suppliers, new circuits, new people. Therefore, it is evident that the whole change process has to be carefully watched and monitored – hence one of the purposes of this Guide.

Critical Quality Indicators:

Most manufacturing organisation have Quality Measures which indicate all aspects of Quality. It is these that must be scrutinized to view the supply chain.

THE EFFECT OF MERGERS, RATIONALISATION, AND RESTRUCTURATION ON THE SUPPLY CHAIN

The main reasons for mergers and take-overs is to make the new company more efficient than either of the two were before. Generally, synergies between the two Research & Development functions, and between the two marketing functions provide extra competitive strength.

The other reason is the rationalisation of administrative, production, and distribution functions. Few mergers indeed have occured in the last years without incurring a 10 to 15 % reduction in production staff and, consequently the closure of manufacturing sites.

Perhaps what is less well known is that the financial benefits accrued from the two companies' combined supply chains can be very large. Especially if, within either or both companies, the supply chain was not well developed. The leverage towards suppliers, transporters, and distributors, becomes massive – in fact the increase is exponential. At the same time the complexity also becomes exponentially bigger. This gives a great challenge for supply chain strategy and planning methodology.

The merger situation itself gives the opportunity to improve radically, or recreate entirely the supply chain.

It should however, be realised that the supply chains of both companies are highly exposed during and just after a merger. This is due to confusion and misunderstanding which could occur between the two companies, especially so if their policies and systems were very different. External relationships could also be significantly affected. Careful management, foresight, and clear communication are required in order to reduce problems to a minimum.

Furthermore the destabilising effect on staff could induce an array of quality problems, unless the situation is well foreseen and well handled in practice. For instance, when production, QC, or warehousing, or any other staff for that matter, feel threatened in their job, potential mistakes, intended or accidental, might occur. This could be caused by lack of attention, lack of care, or lack of motivation. There have been occurences of sabotage for these very same reasons.

Therefore, in any merger or rationalisation situation, a heightened degree of vigilance by management is necessary. Particular attention must be paid to stock levels. There is a danger that strikes and other industrial actions might take place. This is potentially the most dangerous and the most serious quality / supply chain issue in any merger or take over.

Nevertheless, the expected benefits will be derived from the reorganisation of the chain. High quality communication is a key necessity. This, coupled with fast and measured decision making, should be paramount in order to assure staff of their future – even if their future is not within the remaining company structure. But this is a much wider subject than the supply chain and a lot of these considerations are equally true when rationalisation and restructuring take place within the same group.

A further phenomenon is that of resentment. It often happens and can create many problems. Some of these can affect the supply chain when a product is transferred from one factory to another and the first deliveries arrive. In this situation, there is a natural tendency to examine and analyse the product with undue scrutiny. Rejections for minor cosmetic faults have been known. For these and other reasons, the relationship between the two factories could easily become strained.

We may take the example of a suppository pack. Marketing required a large amount of carton on which adequate text could be written around each individual suppository and which also looked very attractive. The pack had been developed specially for marketing reasons.

When another country launched this product, they refused the pack for spurious ecological reasons, deeming that there was too much wasteful carton and that their clients would object to this. What really motivated this reaction was that the factory making this pack had been allocated some work which had previously been done by the other factory.

When the economics of the change and supply chain were explained to general management, the pack was accepted. In fact, when *their* Marketing people saw it, they liked it. Large amount of energy, bad feeling and wasteful investment (for the alternative pack), was incurred.

Technical, process, and packaging differences between one factory and another can similarly create a supply chain problem. An interesting case arose when the cream and ointment of a production was transferred from one factory to another. The formula was the same but the mixing equipment was different. This resulted in the fact that the density of the cream (which was not a critical specification, as it did not affect the efficiency of the cream) was different and hence required a longer tube. This in turn required a longer carton. Both factors had to be approved by the local marketing, because customers were going to get a longer tube in a longer box. But, above all were going to have a less dense or "lighter" cream. There was very little choice and everybody agreed.

However to obtain marketing agreement, new artwork for tubes and cartons, plus their making and printing, took 4 months. During this time an out of stock situation could easily have arisen. Worse still, a major quality problem due to tubes being overfilled could have ensued. Fortunately there was a reserve stock.

The obvious lesson from this example is that in order not to jeopardise quality and the supply chain, stocks should be sufficiently high to buffer any technical, marketing, or social situation during the critical transfer phase. Nonetheless, systems must be in place to reduce this high stock level situation as soon as it is practicable.

The major aspects concerned in a merger or rationalisation are enumerated below.

- Suppliers change
- Sourcing of finished packs change
- Different stock policy
-      »   »   cycle times
- Transfer of production issues   – technical
                                         – process
                                         – packing

- Different quality standards
- Change in artwork
- Cycle times
- Lead times
- MRP or other planning system differences
- Transport methods
- Palletisation »
- Labelling   »
- Coding   »

Different contract arrangements   – Short term
                                         – Long term
notably with   – Distributors
                                         – Wholesales
                                         – Transporters
                                         – Suppliers

IT System differences

Relating to critical quality issues, the following must be noted:
- Transfer of equipment – requiring re-validation
- Change of equipment- requiring re-registration
- Transfer of process know how   – Transfer of staff
- Comparison of training programmes

Critical Quality Measures:

- Comparative and new service levels
- » » client satisfaction indices
- » » cycle times – Secondary
- » » cycle times – Primary
- » » lead times
- » » stock levels – Value
- » » » – Months sales
- » » cost of distribution
- » » method of dealing with complaints

Outsourcing

Outsourcing means putting some part of an activity of a firm to an outside firm. It is generally done with activities which are considered as non-core activities of the firm's business.

It is generally done in order to:
- save money,
- being able to concentrate resources, especially management resources which could be channelled to other activities,
- save space,
- have a better control over fluctuating volumes of activities,
- have better service and quality of the outsourced activity,
- reduce capital investment.

In the pharmaceutical business, outsourcing has been fairly behind other industries. This is mainly due to lesser pressure on costs, and many myths.

The first thing that the industry started to outsource was in services such as catering and cleaning. Nonetheless, cleaning in certain manufacturing areas should be considered as core competence – the cleaning of and maintaining of sterile rooms, is a case in point. Travelling arrangements, heavy equipment maintenance, service equipment maintenance, specialised equipment maintenance (such as balances and analytical instruments) are other examples where outsourcing is widely practised within the industry.

Generally speaking service activities can be more readily outsourced without impinging on the core activities of a firm. Recruitment, Information Technology, Training, Payroll are other activities often seen to be outsourced.

As far as the supply chain is concerned all suppliers, transporters, deliverers' work, are in fact outsourced. So all the control mechanism and contract arrangements which apply to these should apply to any other outsourcing activity.

The traps and difficulties are generally to find the suitable company or partner to whom one outsources.

Often, in order to get the best quality and service, partners have to be trained by the outsourcing firm, although this can be costly and time consuming and many of the weaknesses of in-house sourcing could be transferred.

In any event, the quality and service of the outsourced activity is the most critical aspect to be aware of. Any decision on outsourcing has to be weighed in terms of advantages versus control, training and sustained outsourcing management costs.

The basic theory of outsourcing is that giving a part of one's activity to someone else, who does nothing else, whose core activity it then becomes, can do it more efficiently, as he has only *that* to concentrate upon. It is very similar in concept to dividing a large company into Business Units, where more concentrated attention can be given to a smaller part of the business.

Partnering, Trust, Loyalty are essential and are cost effective features of a successful outsourcing operation.

It is very important to make sure that alternative partners exist and/or that re-insourcing i.e. the taking back of an activity which has been outsourced, can be effected. Pressure on quality, service flexibility, and especially cost, can thus be maintained.

Outsourcing of the manufacturing and/or packing process is now coming to be practised in the industry. As mentioned before, this is much more widely practised in other industries.

Generally, older and cheaper products are outsourced by some pharmaceutical firms. Provided that specifications are clear, quality assurance conforms, and that partnering arrangements are good, there is no problem to outsourcing. Management time and competence to outsourcing must never be underestimated.

16 | *MANAGEMENT AND THE SUPPLY CHAIN*

**Overall Supply Chain Efficiency Measures
in the Pharmaceutical Manufacturing Industry**

The following table (Figure 16.1) shows some of the main quantifiable measures which can be used to monitor the efficiency of the supply chain. They reflect the outcome of the way the chain is managed in financial terms and also in terms of customer focus.

Companies define these measures in different ways. There are no tight rules as to how to define these, however some of these measures require comments.

The main point is the way in which these measures fluctuate over time. Whilst it is normal that some fluctuations do occur, it is the wild and unpredictable fluctuations which might give rise to concern and which could impinge on quality.

a) *Days cover of stock:*

This is expressed in days or average number of days of sales. If sales vary or go up fast or go down fast, then the stock cover figures do not mean a lot.

They can be expresed in unit boxes or in dollars.

It is important to know whether the figures relate to:
– all products,
– one product (molecule),
– one product line.

Probably for expensive products, all of these figures are necessary, but the mixing of figures always presents a risk in interpretation. If a sales campaign is being prepared, then again figures (either before, during, and shortly after the campaign) mean little.

MANAGEMENT AND THE SUPPLY CHAIN

Some Overall Supply Chain Efficiency Measures
in the Pharmaceutical Industry

FORECAST	STOCKS	MANUFACTURE	DELIVERY	COSTS	CUSTOMER
d) Accuracy	a) Days cover of stock : - raw materials - overall - work in progress - finished goods c) - Volume of stocks	- Lead times - Cycle times h) - Production plan performance i) j) & k) - Number of pack changes/year - Number of pack changes as a percentage of total number of lines - Average time for pack changes - Reject levels - Right first time - Number of batches reworked - minimum/maximum order level f luctuation - Flexibility	e) - Accuracy f) - Back orders - Out-of-stock : . number of days . number of items . values - Number of rush orders	b) - Value of stock g) - Out-of-stock value - Distribution costs - Cost of rework - Cost of losses - Cash recovery time : . from delivery . from order	- Complaints : . quality . services . delivery

Figure 16.1

Seasonal variations must be taken into account. An antibiotic might sell three times as fast in winter than in summer. Therefore, just before the winter, its stock must be three times as high as it was before. A flu vaccine stock may be 500 times higher in autumn (in the northern hemisphere) than at any other time of the year. Antiasthmatic drugs also generally have two peaks a year.

b) & c) *Value and Volume of Stock:*

It is necessary to understand that the objectives of an efficient supply chain management are financial. "Financial" does not mean necessarily only cost. It means cost advantages and also competitive advantages, including customer satisfaction and service levels. When one has a very expensive product, one would wish to have low stocks. This is, providing one does not lose sales as a consequence.

Similarly, if one has a relatively low cost product, one does not object to having a larger volume of stock. This is why it is necessary to differentiate and understand figures concerning different products.

d) *Forecast accuracy:*

Forecast accuracy is generally expressed in volumes in relation to forecasts expressed at a certain time before sales. The time might be two or three months. Again, it is important that this time period be constant.

e) *Delivery accuracy:*

Delivery accuracy is the acid test in the supply chain. It reflects the service levels. Sometimes it is referred to as "In-market line fill".

It is generally expressed by a percentage of orders delivered in the period requested to the customer (within, plus or minus, an agreed number of days).

f) *Back orders:*

Back orders are either the number of:
– items,
– packs,
– product groups,
or their total or individual values which are not delivered on the agreed dates.

A further refinement of this indicator is the length of time the item is late (see below).

g) *Out-of-Stock:*

Out-of-Stock expressed in days, by number of items, or in value.

It is, perhaps, here necessary to state that in the pharmaceutical industry, the value in financial terms of "out-of-stock items" is very relative. When a patient needs a drug, the fact that it costs $50 or $1 has secondary importance to its availability.

The author once heard the technical director of a large company say that such and such an item was out-of-stock and that it did not matter because it cost only $2.50 a box. He forgot that it was the only product for a particular ailment, and that a number of patients virtually could not live without it. Fortunately, the aforementioned technical director did not stay too long in the company and thus did not bring this type of culture into the organization.

h) *Production plan performance:*

This is the actual number of units produced compared with the planned production during a month or a given period of time.

This is not a very important measure, unless there are big deviations. Deviations may be due to production problems, or sales figure changes, either upward or downward.

Consequently, this figure must always be compared with the Forecast Accuracy figures.

i) j) & k) *Number of Pack changes and the Average time to change:*

The number of pack changes per year of current production (i.e. not new products), is important to know. This is because it often reflects a *new* supply chain. It is also useful to know what this represents out of the total number of lines.

Sometimes marketing wish to change packs two or three times a year without realising the full supply chain implications, nor the full cost of the changes. In the cosmetic and O.T.C. industry, pack changes are much more frequent and thus their cost is much better apprehended. But, in this case, the supply chain is adapted to frequent changes, stocks are generally lower, and there is a tighter control on those items which are to be replaced.

The average time to change from one pack to another gives a very good indication of the flexibility and adaptability of the company. This figure is generally measured either as absolute time, or as a percentage deviating from the agreed times.

The importance of these figures and measures is not their absolute meaning at a given moment in time but:
– their trends,
– their potential misinterpretation by some people,
– their definition not being changed.

These measure are not directly related to quality. However, when one assesses quality in a company's or a factory's supply chain – these figures give a very good guideline to supply chain management. So if there were important quality issues, they would be perceived soon enough. It is also important to understand these terms precisely, in order to know the way they are used in a particular company.

Conflicting Objectives within the supply chain

One of the main challenges of the supply chain is that of the harmonising of and the aligning of objectives. Many of these objectives can be contradictory.

For instance, choosing and arbitration might not be easy between high stocks and long economic production runs. Similarly, the choice between satisfying two clients requiring the same product (made on the same line) might not be easy, especially if there are no financial advantages between the two choices.

Herewith is a list of typical supply chain Dilemnas or oppositions in objectives:
- Line change over times versus production times,
- Staff allocation between two production lines,
- Regulatory requiring procedure A or B,
- Use of limited stock for two competing orders,
- Cost of information system versus advantages from I.T.
- Cost/service between one supplier an another.

Whenever financial considerations favour one decision against another – the problem is relatively simple. However, the financial considerations are not always decisive.

An example: a manufacturer had the choice between two suppliers.

Supplier A was 20 minutes away from the manufacturer. Supplier B was 300 miles away, but quoted 20 % cheaper than A. The nearer company was chosen as specifications often had to change – and very rapidly at that. Nonetheless, this agile management technique was difficult to explain to the Financial Director!

Whenever one is confronted with diverging objectives, one concentrates one's thoughts on the two objectives in question. This becomes then a twin tunnel vision. Generally, the best way to resolve the dilemma is to try to see the two problems in the context of the whole business.

By putting the objectives in perspective with, and in proportion to the whole supply chain, and in the overall interest of the company, one can often arrive at the right solution relatively simply and quickly.

Proximity Management and the Supply Chain

Speed of decisions is obviously important for supply chain management (or any good management for that matter). It is part of agile techniques. (see section on Agile Decision Taking, Chapter 1).

One has often come across a situation when a production line or process was stopped due to an apparent problem. The problem, in the absence of the department head, was referred to Quality Assurance, engineering, and even to the factory manager. Whereas the department head could have solved the problem within 5 minutes – without having to stop the line.

Even with the best of intentions, in the best managed factories, department heads are not always there. But the fact has to be realised that his or her absence, could affect the supply chain. In his absence, the information circuits have not only to be clearly defined and known by the remaining staff, but the speed of decision making must also be clearly understood and applied.

A factory manager or supply chain manager must know where key and rapid decisions could be taken.

Quality Assurance and Quality Control are an obvious area where speed of decision is vital. Purchasing, Finance, and production management are equally points where fast decision making can significantly affect the supply chain.

Supply Chain Culture

"Culture" can be defined as a set of postulates, expectations, or rules accepted by a group of people. It gives a common perception of reality which is the basis of identity for that group in relation to another. Culture in the supply chain is a way of looking at and behaving towards common goals in partnership between two or more companies or departments in the linkage of the chains.

Figure 16.2 attempts to show how this is reached from a simple supplier-manufacturer relationship to a supply chain relationship. This Culture induces a certain common behaviour pattern.

The Supply Chain and the Budget

A certain number of important figures, as we have seen, emanate from the supply chain, these include:
– stock levels,
– lead times,
– purchased items cost,
– service levels,
– etc...

These figures, one way or another, go into the budget.

As with all budgets they should be ambitious but realistic.

The pressure to make these figures optimal is the essence of supply chain management. But once these figures have been optimised, any further pressure to "improve" them can lead to major quality problems. Poorer quality purchased items may have to be substituted, short cuts may be made in cycle times, stock might reach critical levels.

C.E.O.'s who put excessive pressure, especially between budgets, to reduce cost must bear this in mind.

However if undue or unexpected budget cuts are made, the ricochet effect can be dangerous. By the ricochet effect, we mean that at the next budget exercise, less ambitious figures may be given under the assumption that, later in the year, cuts or pressure will again be exercised. This is a culture where "safe" budgets are made which denotes firstly, a lack of Trust, and secondly, a lack of transparency: all essential features of good supply chain management. Furthermore, it reduces **ambition** and **reach** which is necessary in a modern competitive organisation.

But the perversity does not stop there, it cascades all along the organisation. Because anyone making any part of the budget will adopt similar attitudes. It furthermore undermines trust, which is one of the basic necessities in the pharmaceutical industry and is recognised as a major management concept of the future (see section on Trust, page 184).

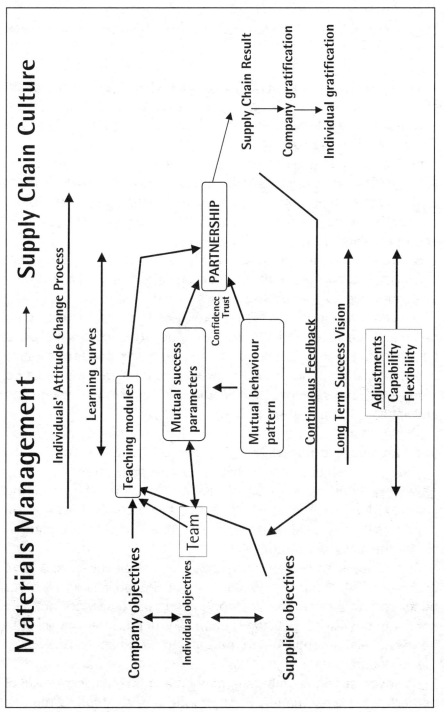

Figure 16.2

Supply Chain as a Management Attitude, Teaching, and Quality Tool (Lever)

Understanding, managing, and improving the supply chain necessitates a complete and original way of looking on the manufacturing, technical, and logistical operation in the pharmaceutical industry.

For, by understanding the supply chain, one can understand the operation as a process in its entirety, holistically. This allows all the weak links, problems, barriers, and constraints to be clearly seen. It brings an insight into quality issues which might not have been perceived before.

By wishing to improve continuously, one can identify the effect of constraints and one can prioritise one's action by more clearly evaluating the benefit/accessibility balance.

Consequently, one can restructure (if one so wishes) the entire operation in a different logic from the past. This logic will allow a different management method to operate, giving better qualitative and quantitative performance results. This management must be more agile and reactive than in the past and must include the anticipation of the unpredictable.

Most people working in pharmaceutical factories do not realise that they see and understand only part of the operation.

They can see one step behind and one step ahead, but they have very little further vision of the line, which is often long and complex, leading from the supplier to the client.People work in departments – "the tablet department", "the ointment department" or "the packing department".

Departments have local allegiances, local objectives, and local priorities. They tend not to care about the objectives of other people or other department. Departments, therefore, are not only compartmentalised, but they often surround themselves with barriers. This is especially true if caused by pressure of work or the workload is very high.

However strong a manager might be, the breaking down of barriers is no easy matter. Furthermore, barriers tend to regrow.

Only by removing the ground for barriers i.e. having no separation of departments, objectives, and priorities, can the problem be overcome.

In the past, managers concentrated their efforts on a particular activity. Top management felt that if each individual activity was well managed, then the whole must be well managed. This might have been true when trades and specialised knowledge were absolutely necessary to accomplish tasks. Today, with modern high performing, automated or semi automated equipment, and sophisticated P.C.'s and computers, a much more general knowledge is required.

The famous adage about the stonemason cutting stones which were destined for a wall in cathedral is very relevant. When asked what he was doing, there were at least three possible answers:
  - I am cutting stones,
  - I am making a wall,
  - I am building a cathedral.

It is the latter answer which shows the modern attitude required. Complexity and speed have both increased. Decisions have to be taken faster and by people much lower down the hierarchical scale. They can be taken only if a much wider knowledge of the process exists.

A cultural change is required. This book is not about re-engineering. However one cannot look at and analyse the supply chain without coming to the conclusion that in order for the client to be closer to everyone, a total change in the industry is required. Agile management leading to mass customisation will be perhaps be one of the results.

Conversely, and in parallel by looking at the process in this manner, many quality aspects can be brought to light, which previously were not apparent. Large stocks often covered or hid quality problems, as well as technical ones.

As an example, one factory, was working with stocks of three weeks of work in progress, and six weeks of finished goods. A certain tablet product had regular losses of 4 to 6 %. This slowly crept up to 10 or 12 %. No one particularly cared, because the production was not very expensive and no out-of-stock-situation occurred. It was only when a new department head, who knew about supply chain and analysed the problem – that a quality improvement reduced the losses to 1,5 % and the stocks to two weeks!

The next chapter of this book, is partly devoted to a teaching module aimed at making sure that all levels of staff understand, at least the broad aims, and constraints of the supply chain.

17 SUMMARY AND THE TEACHING OF THE SUPPLY CHAIN AND A LOOK AT THE FUTURE

This last chapter has two purposes, one is to try to summarise the key features of the supply chain, its complexities and dynamics. And, to give a practical tool of what a teaching module could be for the staff of a manufacturing unit in order to understand better the supply chain. Secondly, it glimpses at the future and the possible evolution of some aspects of the industry.

The title of the module is:

"Making the supply chain understood both within and outside the Organisation".

The module summarises the key points of the supply chain.

It starts with a brief introduction of the economic and industrial background to the pharmaceutical industry as seen at present.

It goes on to explain the objectives of the supply chain, and compares the new way of working with the past.

It then goes on to explain the "difficulties" (of the understanding) of the supply chain by showing its wide scope, its complexity, interactivity, and dynamic nature.

It shows a concrete example of Aerosol production, the number of partners, steps, and interaction points involved in a chain.

It explains what transactions are, the dimensions involved, and the constraints within which one has to operate.

It proposes a framework to ease the understanding and application of supply chain problems.

It gives a list of the partners involved in a chain, it shows the need to coincide personnel, company, supplier, and customers' needs and capabilities.

It goes on to give four examples to show the need to coincide objectives and relate interactions.

It concludes by showing the cultural effects of new supply chain thinking.

Also included is a facultative part of the module about the need to identify oneself in a changing industrial environment. An explanation of the need to "unlearn" is also given.

The module takes 35 minutes to present. It should be interactive with the audience, which could take the session up to 1 hour, 1 hour 30 minutes.

LOOKING AT THE FUTURE

There is no harm to dream in a serious management book. Mass customization is a subject which is not far away.

Mass Customization and Pharmacogenetics

Mass customization is not as yet practised in the pharmaceutical industry.

It is therefore difficult to describe it; one can only propose or envisage tentative models. This however allows some freedom in looking forward to the future and being creative.

Mass customization in other industries means making products economically in specific industrial quantities and according to the specifications and the needs of the individual consumers.

One may therefore conjecture, in order of complexity the following scenarios:
   a) more accurate and specific diagnostic procedures, leading to more personalised doses (even of existing medicines).
   b) larger variety of dosages for more accurate targeting of the degree of illness, age, and the weight of the patient.

c) Pharmacogenetics, that is linking patients genotype with his own metabolizing system will provide new drugs, new dosage forms and different drugs regimes and resuscitate old ones. These however are likely to be made in smaller quantities.
d) with evolving gene therapy on specific molecules, more varied and specific delivery forms
e) specific combination of medicines in view of patient's requirements. These medicines need not necessarily come from the same company or source of research and development.
f) devices and appliances for drug administration eg. once a year implants, etc...

One can imagine the patient diagnosing his immediate requirements and by tele transmission informing his prescriber, who then orders the suitable dosages.

One can envisage the doctor prescribing, after tele consultations on a screen relayed to a factory or pharmaceutical manufacturing establishment (this might be mobile), the specific remedy. This remedy is subsequently delivered to the patient.

Specific feedback information from the patient may then modify the regime.

Clinical trials will become far more patient profile specific and indeed many production processes will take on the aspect of Clinical trials manufacture.

Personalised leaflets

Personalised dosage instructions combined with dietary measures, if necessary, would be included in the packaging.

At a further date, a patient's whole specific physiological make-up would be known and monitored, and molecules as well as gene modification would be administrated.

In this scenario, a once a year implantation (like a car going into a garage) would replace the conventional pharmaceutical supply chain.

In parallel with this, it might not be inconceivable that the patient will require a lot of personal psychological attention. He would go to his local pharmacist, homeopath, herborist, or psychopharmacologist to have counsel and individual locally made up preparations – as in the long past.

Other Manufacturing methods

The classical way of tablets for instance, may be superceded by extruded "wire like" forms, where different products can be fused. (Similar to multicoloured tooth paste.)

Printed layers of products could also be imagined to "stick" together, each layer being a different active ingredient (like layers of multicoloured liquorice sweets).

The most important factors in developing and thinking up of new drug or medication forms is to change the MIND SET. This is the challenge of future decades!

MAKING THE SUPPLY CHAIN UNDERSTOOD

WITHIN AND OUTSIDE

THE ORGANISATION

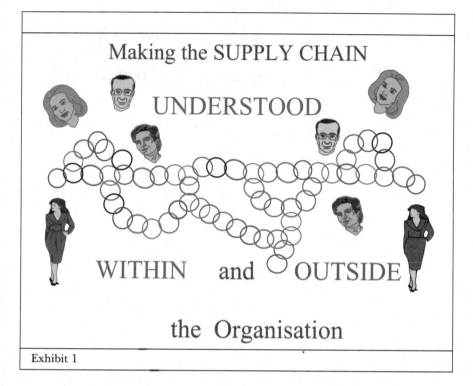

Exhibit 1

1. Understanding the Supply Chain within and outside the organisation:

The purpose of this presentation is straightforward. It is to ensure that by the end of this presentation, everyone in this room will understand the concept of the supply chain, how it functions both internally and externally in an organisation, its complexity, and how it can be managed to the benefit of the organisation. The presentation will include real life examples to illustrate key points and emphasize their importance. The presentation is in no way exhaustive, nor is it intended to be. Rather it is anticipated that it should be a forum for discussion with the audience in which personal experiences can be debated. Thus questions and comments are welcomed throughout.

The Arborescent Nature of the

SUPPLY CHAIN

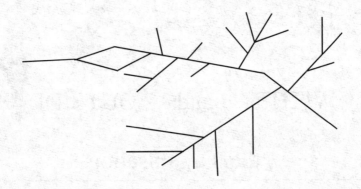

Exhibit 2

2. What is the Supply Chain?

When one hears the word "chain", one thinks immediately of a series of links which are joined together to form a straight length (show example of a chain). Therefore when one thinks of a Supply Chain, it is easy to picture the same simplistic view which illustrates one very important point – that each link in the chain is interconnected with its neighbour and if something disturbs one link, it may have implications for several others. Although this chain (show example) appears already to be complex, unfortunately in reality the supply chain is not even as simple as that. In real life, it is more like this (show example 2, a more complex chain) and this is again a simplistic representation of something that looks more like this (show slide arborescent nature of supply chain). A tree, and let us not forget that trees have roots as well as branches, illustrates very well the supply chain which is complex, not only in its individual parts, contracts arrangements, specifications, cycle times, but as a whole integrated chain. A chain which is never straight or straight forward, but much more like an arborescence or root structure.

Its efficient management also becomes more complex with a large array of interactions, many choices, alternatives, paradoxes, and dilemnas. Decisions have to be taken faster. The consequences of these decisions are more far-reaching and costly. Constraints have become tighter.

Priorities are more difficult to manage.

Because it is more complex – it is more difficult to understand. More difficult to understand by every one along the chain. Yet it is only by understanding the whole chain by every link, and by everyone involved in the chain, that its management can be optimised.

By understanding the Supply Chain, the main objectives of a company or a manufacturing unit can be kept aligned and in focus by every member of the staff.

Let us have a look now at the objectives of the Supply Chain,

---

Objectives
of the Supply Chain (I)

The objectives are that **ALL** that goes in and around a

pharmaceutical product from the **SUPPLIERS** of components

to the finished product and services are

delivered to the **CUSTOMER** in the most **ECONOMIC** and

TIMELY manner, and at the **QUALITY** that the user expects.

Exhibit 3

---

3. Objective of the Supply Chain:

The objective of the Supply Chain is very simple.

The customer should receive the pharmaceutical product when he requires it, at a price which he can afford, and with a level of quality that he expects.

The "customer" in this case is a term not just confined to being the end-patient but is any of the recipients of any item or service which forms part of the process of the pharmaceutical supply chain.

Objectives
of the Supply Chain (II)

In order to achieve this Objective

the INVOLVEMENT and COMMITMENT

of ALL partners involved (Suppliers,

Manufacturers & Customers...)

is ABSOLUTELY ESSENTIAL.

Exhibit 4

4.

In order to achieve this objective, it is paramount to have the full involvement and commitment of all the "partners" in the supply chain – the suppliers, manufacturers, and customers of each mini-segment of the supply chain.

All Benefit from good Supply Chain management

Increase MARGINS

Exhibit 5

5.

The motivating point in this is that all these partners will benefit by increased margins and therefore greater profitability.

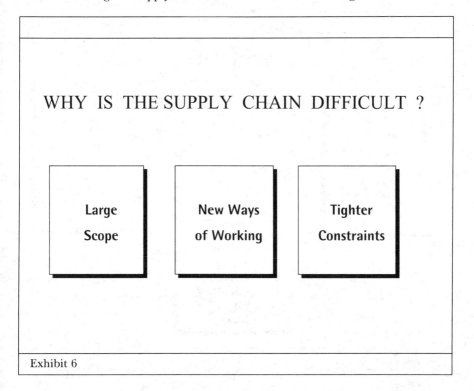

WHY IS THE SUPPLY CHAIN DIFFICULT ?

Large Scope

New Ways of Working

Tighter Constraints

Exhibit 6

6. Why is the Supply Chain difficult?

There are many factors which contribute to the increasing difficulty in managing the supply chain. Broadly speaking they fall into the following three categories: Large scope, New ways of working, and Tighter constraints.

Let us look at each one in a bit more detail

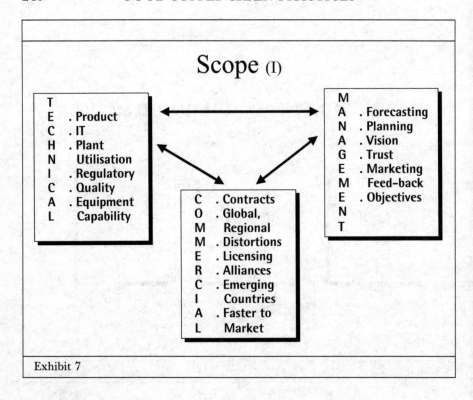

Exhibit 7

7. Scope (I):

The scope of the supply chain is enormous. It encompasses most obviously technical aspects such as (refer to list on transparency). Commercial aspects are equally as important with today's environment of rapidly changing market conditions where new markets are opening up and new competitors are increasing in numbers (refer to list on exhibit). Management factors are gaining more and more recognition in business and are critical to the success of a modern and efficient supply chain. To have reliable forecasts is a must, to have trust in your supply chain partners is an immeasurable benefit, and to have marketing feed back helps not only with planning but with other technical aspects such as equipment capability.

Scope (II)

- Complex
- Interactive
- Dynamic
- Far Reaching

Therefore it is difficult to make the Whole understood,

and difficult to identify Core Competences.

Exhibit 8

---

8. Scope (II):

All these factors, and this list is by no means exhaustive, are in themselves complex – (an item of production equipment for example). But not only are they complex but they also interact with one another dynamically – not just in a static sense. Whenever there is a change in condition of one, it may and can affect a number of the other variables. It is this complexity, this dynamics, and this interactivity, which makes the supply chain difficult to understand.

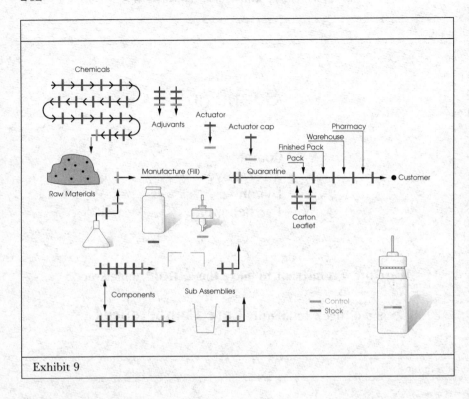

Exhibit 9

9. **Aerosol production process:**

One can illustrate this very well by looking at the manufacturing process of an aerosol inhaler. There are hundreds of little processes which make up the overall manufacturing process, each of which needs to be planned, managed, coordinated, etc... so that the end result is that the customer receives his product when he walks into his local pharmacy. (Elaborate by selecting one or two manufacturing stages on transparency eg. aerosol valve has many components plastic and metal, the supply chain of which has to be managed, eg. some active ingredients are the culmination of 30 stages, the chemical synthesis lasting several months).

BETWEEN each LINK, there is TRANSACTION

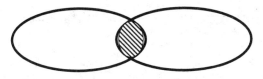

TRANSACTIONS cost MONEY and TIME

Control,			Specifications,
Storage,			Manufacturing,
Registration,	Invoicing,		Tooling,
Computerisation,	Payments,	Coordination,	etc...
Space,	Receipts,	Negotiation,	
Transport / Information,	etc...	Choices,	
Telephone,		Decisions,	
Fax,		Management,	
Packing,		Training,	
etc ...		etc ...	

Exhibit 10

10. Transactions:

Thus all these links in the supply chain are joined together by transactions, either internal or external, to a particular organisation. These transactions are not just the invoices, the payments and the receipts that one might traditionally think of, but also others such as intermediate storage of products or components, negotiations over specifications, etc, etc... All these transactions cost both time and money and this is the reason why the good management of these can lead to substantial savings for all involved.

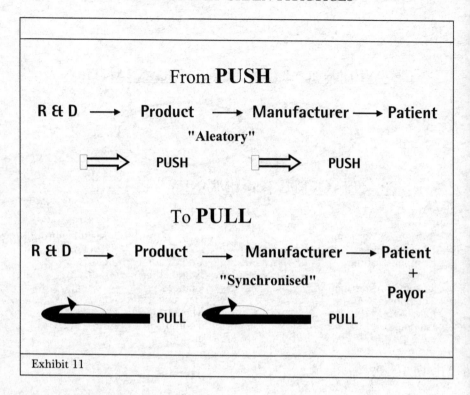

Exhibit 11

11. New Ways of Working:

The effective management of the supply chain will mean that new ways of working must be introduced. In the past one operated a push system where research and development pushed their products to manufacturing who then pushed them to the patients whether they wanted them or not. With the supply chain, it is the patient and/or payor who pulls the product from the manufacturer and ultimately research and development. This is a fundamental change which I will explain in greater detail.

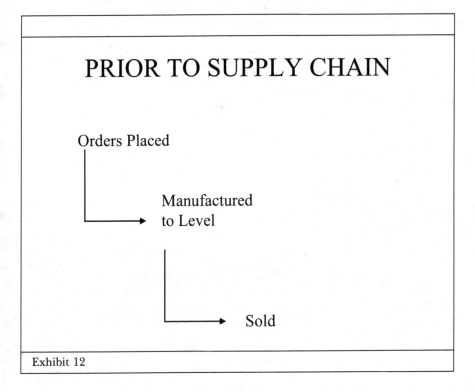

Exhibit 12

12. **Prior to Supply Chain:**

Before the advent of the supply chain, orders were placed by the vendors, the product was then manufactured to fulfill the order requirements and delivered to the vendor who then sold the product (the Push system).

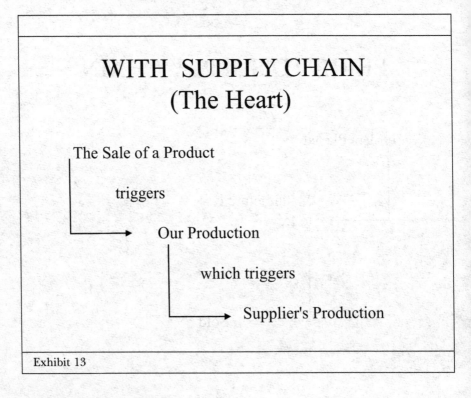

WITH SUPPLY CHAIN
(The Heart)

The Sale of a Product

    triggers

        Our Production

            which triggers

                Supplier's Production

Exhibit 13

13. With the Supply Chain (I):

With the supply chain, it is the sale of the product which triggers (pulls) the pharmaceutical factory to manufacture. In turn it is the consumption of the components by the pharmaceutical factory which triggers the suppliers to produce more.

WITH SUPPLY CHAIN

All the partners along & beside the Chain must know :

. Quantities sold
. Stock level
. Daily Production
. Reactivity

IMMEDIATELY

ALL THE TIME

Ressources have to be managed with more precision.

Exhibit 14

14. **With the Supply Chain** (II):

For this system to work, it is essential that all the involved parties in the supply chain have immediate and continuous access to the following information: quantity sold, stock levels, daily or perhaps hourly production rates, and the reactivity of each "supplier" in the chain (this could be the pharmaceutical manufacturer!).

PREVIOUSLY

One Sold what was <u>Made</u>

TODAY

One Makes what the Market Requires.

Exhibit 15

---

15. Summary:

So in summary, previously one sold what was produced. Now one makes what is required by the market.

> Let us look in detail at the advantages and disadvantages of each system.

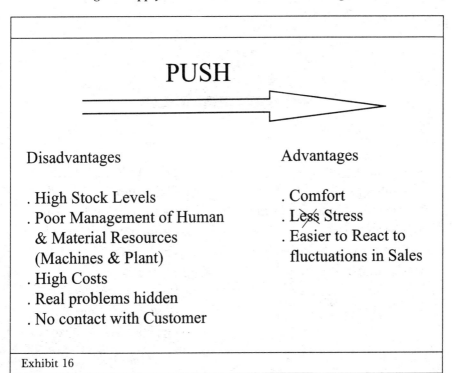

PUSH

Disadvantages	Advantages

Disadvantages

. High Stock Levels
. Poor Management of Human
  & Material Resources
  (Machines & Plant)
. High Costs
. Real problems hidden
. No contact with Customer

Advantages

. Comfort
. Less Stress
. Easier to React to
  fluctuations in Sales

Exhibit 16

16. **PUSH system:**
(self explanatory)

but the real problems are hidden because of high stock levels and lack of urgency.

PULL

Disadvantages	Advantages
. Do it right first time	. Direct Contact with
. No Buffer - Risk ++	Customer
. Continuous need for Orders	. Optimisation Use of
. Higher Stress	Human & Technical
. Information Flow difficult	Resources
to put in place	. Lower Stocks
	. See Sales figures
	(faster reactivity)
	. Greater Autonomy

Exhibit 17

17. **PULL system:**
      self explanatory)

highlight increased stress but greater autonomy and above all visibility.

FROM FUNCTION TO PROCESS

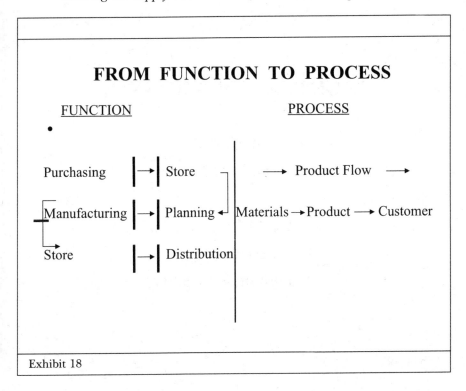

Exhibit 18

18. Function → Process:

The supply chain is therefore all about moving from a functional organisation where each "section" worries only about its own particular speciality without regard to the consequences that its actions might have upstream or downstream, to a process led organisation driven by the flow of the product.

TIGHTER CONSTRAINTS

- Optimal Cost,

- Optimal Speed,

- Lowest Stock

- Lowest Material Loss,

- Fast Reactivity,

- Maximum Flexibility

Exhibit 19

19. Tighter Constraints:

The third category which makes the supply chain ever more difficult to understand and manage concerns the constraints within which one has to operate. These are becoming tighter every day with the need to increase margins, whilst delivering better service and maintaining product quality eg. the drive to operate at an optimum speed for an optimum cost. Note the importance of the word "optimum" which is chosen deliberately instead of "lowest". Another example is the endless pursuit of flexibility and responsiveness to better serve one's customers.

REQUIREMENTS AND CAPABILITIES WITHIN THE SUPPLY CHAIN

Coincidings Needs and Capabilities

- Identify who ALL, my customers, suppliers, partners are.
- What impact my customers needs have on other partners along or beside the Chain.
- Coincide customers' needs with suppliers and partners' capabilities.
- Inform Each about All

Exhibit 20

20. **Requirements and capabilities within the Supply Chain:**

In order to be able to manage the supply chain, one has to be able to understand each link in the chain i.e. each process. In comprehending each process, it is necessary to define clearly one's suppliers and customers, to find out what their objectives and requirements are, and then use these to define one's own objectives and requirements. It is important to quantify what impact each partner's objectives and needs have on the other partners in the supply chain. Then every partner's capabilities must be matched with their respective customers needs, thus creating the integrated and balanced supply chain.

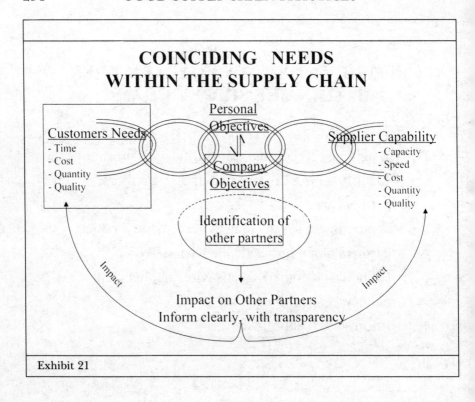

COINCIDING NEEDS WITHIN THE SUPPLY CHAIN

Customers Needs
- Time
- Cost
- Quantity
- Quality

Personal Objectives

Company Objectives

Supplier Capability
- Capacity
- Speed
- Cost
- Quantity
- Quality

Identification of other partners

Impact

Impact

Impact on Other Partners
Inform clearly, with transparency

Exhibit 21

21. **Coinciding needs and capabilities:**

This diagram illustrates the necessity to coincide customer needs with supplier capabilities, and how this shapes both the company's and the individual's objectives. Moreover, it shows how the organisation in the way it operates has a direct impact on both its suppliers and customers. Hence the need for complete transparency – especially if new partners are introduced into the chain.

EXAMPLES OF
SUPPLIERS & PARTNERS

*   Raw Materials Manufacturer
*   Components   "
*   Laboratory
*   Stores
*   Transporter
*   Customs Clearance
*   Ministries (Trade, Health, Finance)
*   Purchasing Dept
*   Finance Dept
*   Regulatory Dept
*   Medical Dept

Exhibit 22

22. Suppliers and Partners:

Here are some examples of suppliers and partners in the supply chain.

Examples in Relating

* U.S. Launches

* Purchasing Quantities and Stores

* Home and Export

Exhibit 23

---

23. Examples:

Follows some examples of coinciding requirements and capabilities which illustrate the need to foresee the impact of changing requirements and capabilities on partners in a supply chain.

The difficulty is to RELATE, to make the connection, to see the relevance, and to prioritise in often conflictual or competing priorities.

These themselves are always difficult, but the importance is to be able to foresee potential conflicts and to avoid them if possible. If that is impossible, then to refer them to senior management as soon as observed.

If we can teach how they can learn to relate and foresee – we have already achieved a lot.

Let us illustrate this by three examples.

Example 1: U.S. launch

A product has been scheduled to be launched in March. Everything in the European factory, which is to supply the product, has been planned with this in view. Similar products are already on other markets, therefore equipment scheduling has to be fairly tight. Suddenly, in December, the factory is told that the product could be launched earlier – in January. This is because FDA are likely to give approval earlier than anticipated. The factory has to reschedule, involving double shift labour. This costs more, yet it is the factory's objective to keep costs to budget!

A dilemma! – only by all staff fully understanding, that 2 months extra sales would cover X times the cost of the extra labour. So this is the Human and Technical aspect dealt with. But what about the Financial?
The extra revenue will go to the U.S. branch and the extra cost of labour on the European branch. Although the firms' overall situation is positive – the European finance is not happy!
A second dilemma! Creative accounting. Cross charging can probably solve the problem.

Example 2: Purchasing Quantities and Stores

A particularly bulky, but inexpensive item, was always purchased in lots of 20 000. Storage space was calculated accordingly. Suddenly the purchasing department finds they could get a 20 % lower price if the orders were in 100 000 lots.
If they understand the Supply Chain, and relate this problem to the lack of space problem in the stores, they, with the storemen, could find a solution. If on the other hand, they order the large quantity – chaos in the stores!
There could be an extra cost of storage outside the premises outweighing the 20 % purchasing cost advantages.
Knowing perfectly the next links (or his customer's) constraints, objectives and requirements, and relating these to his own objectives and requirements and foreseeing and preventing conflicts, are the keys to manage the chain in an agile manner.

Example 3: Home and Export

Home and Export requirements for a particular product fluctuate a lot. Suddenly there is a surge for both. The manufacturing equipment capacity is OK, but one of the component suppliers has problems in meeting delivery dates.Which country to supply? How much to supply – dilemna, conflict of interests, and competing priorities!

This problem is more complex and more strategic in the sense that with a variably selling product, and having to supply a number of markets, probably more strategic stocks or higher stock levels of components should have been kept.

Or a second supplier should be available.

Or in a global company, is there a back up plant?

Again these factors must be understood in order to try to avoid the problems.

In a case like this, final decisions must be referred to senior management.

These examples have been relatively simple and straight forward – cause and effect on one or two products, or on one or two markets, or on a particular department or function.

Most Supply Chain problems are much more complex where we have multiple and multiplying effects, snowball effects, etc...

It is the understanding of these which requires foresight, customer knowledge, process knowledge, and knowledge of linkages.

> Example of this may occur as a result of:
> – machine breakdown,
> – faulty manufacture,
> – unforeseen marketing trend (involving total capacity and therefore further investment or disinvestment issues),
> – supplier breakdown,
> – unexpected regulatory constraints,
> – I.R.
> and many other issues.

One factory had a problem with a tablet press – blackspots. It took about 10 days to solve the problem, during which time three different products had to be made. By the time they were subsequently made a bottle neck occurred in the Quality Control lab which worked overtime in order to release the three products. But a control of raw material for an antibiotic was delayed – so the sterile facility had to stop for 3 days because it was not worthwhile to clean down for another product.

Had the effect of the tablet problem been assessed, not only vis a vis stocks and the market, but also vis a vis the Quality Control lab – the problems could have been avoided.

Insufficient Relating, Process knowledge, and Interaction knowledge!

CULTURAL EFFECT OF THE
SUPPLY CHAIN (1)

Identification

Traditional <u>Crafts</u> ⟶ modified new <u>skills</u>

⟶ polycompetencies

creates

Loss of Reference Points

Loss of Recognition & Self Esteem

Degree of "Importance" not known.

Exhibit 24

---

24. Cultural Effect of the Supply Chain (I):

The transition from a traditional to a supply chain driven organisation obviously necessitates an enormous number of physical changes within an organisation (eg. new planning systems, new organisations, new ways of working, etc...).

However it is often found that an individual experiences a loss of identity when he moves from traditional roles and functional tasks to a broader role where new skills are developed and a wide range of competencies are needed. This is probably due to the uncertainties of the new role – reference points are lost, perhaps there is a loss of recognition or maybe the person just does not have a feeling of his "importance" in the new organisation.

CULTURAL EFFECT OF THE SUPPLY CHAIN (II)

Identification

Everyone has to be able to

IDENTIFY

himself with his work

surroundings

family, friends, etc...

in order to be Recognised and to feel to Belong

Exhibit 25

25. Cultural Effect of the Supply Chain (II):

This is a natural human reaction. Everyone has to be able to identify himself not only at work, but at home and with friends, family, etc... so that he can be recognized and therefore feel that he belongs to a specific social or work environment.

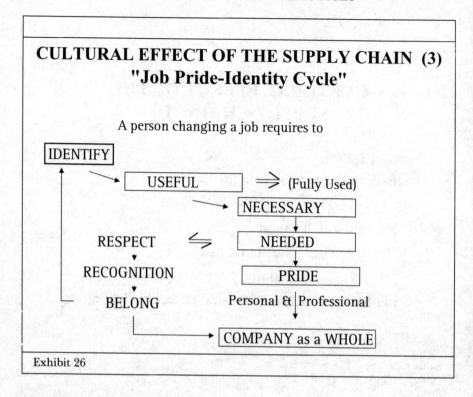

CULTURAL EFFECT OF THE SUPPLY CHAIN (3)
"Job Pride-Identity Cycle"

A person changing a job requires to

IDENTIFY

USEFUL ⟹ (Fully Used)

NECESSARY

RESPECT ⇆ NEEDED

RECOGNITION

PRIDE

BELONG   Personal & Professional

COMPANY as a WHOLE

Exhibit 26

26. Job Pride – Identity cycle:

Thus to put it into a work context, it is important that every person changing job should be able to identify himself and his usefulness in the new organisation. He needs to be shown that his presence is necessary, that his skills are wanted, and that his opinions are respected and recognized. If this transpires then the person will develop a sense of belonging and pride both in himself and in the new organisation. This will of course benefit the company and the supply chain as a whole. If the person does not receive this type of feedback and support, then it is likely that he will only pay lip-service to the new "process flow led" organisation and not be wholeheartedly involved.

CULTURAL EFFECT OF THE SUPPLY CHAIN (4)

Attitudes & Behaviour

. Win Win - No losers

. Transparency & Trust

. Loyalty

. Openess

. Pro-active and Permanent Communication
   with <u>All</u> partners Within and Outside

. Unlearn Functional Thinking

Exhibit 27

---

27. Attitudes and behaviour:

The new organisation must also adopt new attitudes and behaviour. There must be (refer to transparency)
— win win culture
— etc...

The need to "unlearn" or change one's mindset is crucial.

UNLEARNING (1)

- Old methods of working have to be unlearned.

- Functional Thinking $\longrightarrow$ Flow Thinking

- Departmentalisation $\longrightarrow$ Integrated Multi Service

- Compliant Culture $\longrightarrow$ Questioning Culture

Exhibit 28

28. Unlearning (I):

One has to change. One has to forget the "old" ways of doing things – to change from functional thinking to flow thinking, from departmentalisation to an integrated multifunctional team, and from a compliant to a questioning culture.

UNLEARNING (II)

- All parts of the organisation have to unlearn and relearn.
- Relationships and attitudes change :
    - . Win win negotiations,
    - . Transparency,
    - . Dialogue,
    - . Trust, loyalty,
    - . Lack of hierarchical inhibition

Exhibit 29

29. Unlearning (II):

This is applicable to all parts of the organisation so that a clear message is sent with regard to trust, loyalty, transparency, etc...

CONCLUSION

- Understanding Customers needs in terms of Human, Time, Financial, Commercial, Technical, Legal, Space, Other

- Indentifying all Suppliers / Partners, and their CAPABILITIES

- Align objectives of Customers with Suppliers & Company

- Demystify Complexity by Understanding & Relating

- Unlearn and Relearn

- Ensure new job content gets Recognition

- Motivate understanding by ensuring Recognition of New knowledge gained by staff

Annex I

THE HASCOET PARADOX

Hascoët has pointed out that pharmaceutical products are formulated, manufactured, packaged and stored with a view to have the maximum length of shelf life. Most products have a shelf life of between 2 and 4 years. However, from a supply chain perspective, the shorter the shelf life is, the quicker the product has to reach the customer in order to be consumed. Therefore, it might be financially more advantageous for the company producing the drug, to have as short a shelf life as possible.

This is potent reasoning, however the longer shelf life gives a lot more flexibility, both for the user and the manufacturer. It is a buffer which has great practical value. Its financial value however has never been assessed.

The shortest supply chain that this author has known, concerned a veterinary product which had to be injected into cows within 2 days of being made. It was also seasonal i.e. it had to be given in one particular month of the year. Millions of vials had to be manufactured and packed only once a year for a very limited period of time. Three shifts, plus work on Saturdays and Sundays had to be put in place. The product was then dispatched in small vans directly to farms, sometimes motor-cycles or even cycles were used. It was one of the more colourful aspects of the supply chain in pharmaceutical manufacturing.

Annex II

CULTURE CHANGES
necessary in the Pharmaceutical Industry

Passive Compliance Culture → **Transmuting or Post Compliance Culture**

Passive Compliance Culture	Transmuting or Post Compliance Culture
. Non Questioning	. Challenging
. Little and slow Innovation	. Encouraged Innovation
. Lack of Initiative	. Reasoned Initiative
. Long standing Practices	. Opportunity of Creativity
. No Trust	. Measured Trust
. Complex Circuits	. Simpler Circuits
. Administrative Mind set	. Process Mind set
. Procedural)	(. Systematic
. Pedantic)	(. Precise
. Controlling)	(. Reliable
	(. Attentive
	(. Exigent
. Immobile	. Improving
. Fear	. Prudent Courage

INDEX

A

B

C

Printed by Hérissey
at Évreux (France)
May 1998 - 79881
ISBN : 2-906016-01-0-2